KT-481-448

STAY YOUNG
THE MELATONIN WAY

STAY YOUNG THE MELATONIN WAY

The Natural Plan for Better Sex, Better Sleep, Better Health, and Longer Life

STEVEN J. BOCK, M.D., and Michael Boyette

A LYNN SONBERG BOOK

A DUTTON BOOK

PUBLISHER'S NOTE: The ideas, procedures, and suggestions contained in this book are not intended as a substitute for consulting with your physician. All matters regarding your health require medical supervision.

DUTTON
Published by the Penguin Group
Penguin Books USA Inc., 375 Hudson Street,
New York, New York 10014, U.S.A.
Penguin Books Ltd, 27 Wrights Lane, London W8 5TZ, England
Penguin Books Australia Ltd, Ringwood, Victoria, Australia
Penguin Books Canada Ltd, 10 Alcorn Avenue,
Toronto, Ontario, Canada M4V 3B2
Penguin Books (N.Z.) Ltd, 182–190 Wairau Road, Auckland 10, New Zealand

Penguin Books Ltd, Registered Offices: Harmondsworth, Middlesex, England

First published by Dutton, an imprint of Dutton Signet,
A division of Penguin Books USA Inc.
Distributed in Canada by McClelland & Stewart Inc.

First Printing, June, 1995
10 9 8 7 6 5 4 3 2 1

LIBRARY OF CONGRESS CATALOGING-IN-PUBLICATION DATA

Bock, Steven J.
Stay young the melatonin way : the natural plan for better sex, better sleep, better health, and longer life / Steven J. Bock and Michael Boyette.
p. m.
"A Lynn Sonberg book."
ISBN 0-525-94115-0
1. Melatonin—Physiological effect. I. Boyette, Michael.
II. Title.
QP572.M44B63 1995
612.4'92—dc20 95-18652
CIP

Printed in the United States of America
Designed by Eve L. Kirch

This book is printed on acid-free paper. ∞

Contents

Preface

You may not have heard much about melatonin yet—but you will. This natural substance, produced in minute amounts by a gland deep within our brains, promises to reveal deep secrets about how the human body functions.

In my practice, I've been using melatonin for several years. As physicians and researchers delve more deeply into melatonin, they're learning a truth that has long been embraced by Eastern and natural practitioners: Good health is a matter of balance and rhythm.

Many of the problems I see can be traced to a lack of balance in the body. For example, a patient may see me because she is consistently tired in the middle of the day. Maybe she's drinking lots of coffee to try to compensate. Often we'll find that the best medicine is simply to allow the body to heal itself—that once we remove the caffeine and other artificial stimulants, the body finds and restores its own inner balance.

As the most current research shows—and as I've seen among my own patients—melatonin appears to be an essential key to the body's ability to maintain that balance. Melatonin does many things in the human body; in one way or another, all of these functions are involved with maintaining balance. Like an orchestra conductor, melatonin keeps diverse body systems synchronized and functioning as a harmonious whole. It helps these systems communicate with one another, and with the outside environment. It helps repair cells and systems damaged by exposure to toxins and the stresses of daily life. In short, it keeps the vast and complex system functioning smoothly and in harmony.

Scientists are seeing new and elegant relationships between the body and its environment. For example, they are learning that we are less prone to infection and heart disease if we pay attention to rhythms of light and darkness. They are learning that such seemingly unrelated health problems as Alzheimer's disease and diabetes may in fact be related to how we live in relationship to the natural world. And as we'll see, they are beginning to realize that human disease and aging can only be understood if we look at them in the larger context of nature's own cycles of life.

Research into melatonin is revealing new, practical ways to improve our lives and health. And best of all, most of them are simple, natural and low- or no-cost.

In this book, we've focused primarily on strategies designed to optimize your body's *own production of melatonin*. Though we'll also look at the use of melatonin supplements, my opinion is that supplements should be viewed as just that—*supplements*. In other words, I prefer an approach that looks first at natural methods to boost melatonin levels, with supplements used as an adjunct for specific needs.

Some may consider this approach as a little too cautious; the research to date suggests that melatonin supplements are both safe and effective. I believe that they are, and I recommend them with confidence. However, I also believe that the path to good health always begins by supporting the body's own processes. In the pages that follow, you'll find dozens of techniques to take advantage of your body's own natural rhythms, allowing you to safely, simply, and effectively *Stay Young the Melatonin Way.*

STEVEN J. BOCK, M.D.

Part One

THE NEW MIRACLE HORMONE

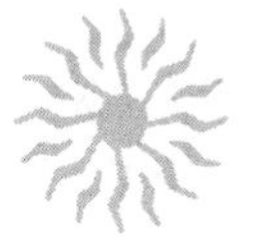

CHAPTER 1

Unlocking the Mysteries of Aging

Poor Ponce de León. The Spanish explorer and his followers were searching for the Fountain of Youth in the New World. Instead, all they found was overgrown vegetation, for which they named the land Florida. Frustrated in his search for eternal youth, Ponce de León returned to Spain, where he grew old and died.

He was neither the first nor the last to ponder the mysteries of aging, or to seek ways of defeating it. From the opening chapter of Genesis, people have tried to understand why we grow old and die, and to outwit (or at least postpone) death's inevitability. Throughout history and in every culture, folklore is full of potions and prescriptions for long life and a healthy, vital old age. In the Caucasus mountains of eastern Europe, legends tout the life-giving benefits of yogurt and goat cheese. In India, yogis practice meditation and drink their own urine. In America, attempts to slow the aging process involve everything from cross-training to crystals. Beginning in the

1800s, millions have flocked to the lands that Ponce de León explored, hoping and believing that Florida's sunny winters might improve their health and somehow slow nature's clock.

Is this annual migration nothing more than a response to vacation ads and real estate pitches, or is there a deeper, more primitive urge at work?

That idea isn't as far-fetched as you might think. For scientists are now discovering that patterns of daylight and darkness affect our health in surprising ways. New findings reveal that daily light cycles regulate basic biological rhythms in animals—making birds restless in the fall and sows fertile in the spring—and humans.

Today, hundreds of researchers, working in university- and government-sponsored laboratories throughout the world, are amassing a growing body of evidence that these patterns of light and dark—and our bodies' chemical responses to them—may hold the key to unlocking the mysteries of aging.

Finding the Fountain of Youth

If there is indeed a Fountain of Youth, it's probably located right between your ears: a tiny conical gland at the center of your brain known as the *pineal gland*. As fountains go, the pineal gland isn't very productive: It releases almost undetectably minute quantities of a substance known as *melatonin* into the bloodstream. But as researchers have learned more about this mysterious and elusive hormone, they have discovered that it has far-reaching effects on some of our most basic bodily processes. In fact, drop for drop, melatonin may be one of the most powerful hormones in the body.

Connected by a direct nerve pathway to the eyes, the pineal gland produces melatonin when darkness falls, helping to regulate the basic daily rhythms of your body. The initial clinical studies of melatonin focused on problems related to sleep and daily cycles; for example, those that affected travelers and shift workers. But as researchers began to look more closely at melatonin, they found that it also has *long-term* effects on the body. And they are now learning that it may be useful to treat a variety of diseases that, at first glance, seem unrelated. Here's why:

One of the most common sources of cellular damage comes from a chemical process called *oxidation*. In common everyday experience, oxidation causes iron to rust, paint to fade, and oil to go rancid. At the cellular level, it causes damage by breaking down the complex and delicate chemical compounds that are necessary to life and health. These chemical attacks can cause a range of health problems, from wrinkled skin to heart disease. And by damaging the DNA of cells, this process can promote cancer by triggering the transformation of a healthy cell into a cancerous one.

It turns out that melatonin may be useful for preventing and treating such diseases because it is one of the most powerful *antioxidants* ever discovered. When it's present in cells, it prevents the chemical damage of oxidation from occurring. By blocking cellular oxidation, melatonin may help prevent the changes in blood vessels that lead to hypertension and heart attacks, and may reduce the likelihood of certain kinds of cancer. (In fact, clinical trials are already under way to study its effectiveness in preventing and treating breast cancer.)

What's more, the discovery of melatonin's antioxidant properties has led to a new way of thinking about aging. Scientists have recognized for years that these kinds of health problems are intimately connected with the aging process, but until now they didn't know why. As researchers learn more about melatonin's role in preventing cellular damage throughout the body, many of them are moving toward a radically new view: *They believe that many, if not most, of such age-related health problems are caused by declining levels of melatonin.* Natural levels of melatonin decrease as we get older, and it appears that this decline may leave our bodies less able to prevent and repair damage caused by oxidation.

So beyond all of these specific health benefits of melatonin lies the biggest promise of all: It appears that *melatonin and the pineal gland control the aging clock itself, and that we can use melatonin to slow down the process.*

That is more than blue-sky speculation. Animal studies have already shown a powerful link between levels of melatonin in the blood and clinical signs of old age. In fact, these studies have revealed that it is possible to influence the aging process—and even extend the "normal" life span of lab animals—by manipulating the animals' production of melatonin. Already these researchers have used pineal gland transplants to extend lab animals' life spans by as much as 25 percent. (In humans, that would mean you could expect to be healthy and active until your hundredth birthday—or even beyond!)

In the chapters that follow, we'll describe these and

other experiments, as well as the latest research on melatonin's role in human development and aging. We'll show you the incredibly vast range of benefits that scientists are discovering for melatonin, including:

- its use as a safe and natural sleeping pill
- its ability to help ease symptoms of premenstrual syndrome
- its potential for stimulating the immune system and reducing the likelihood of infection
- its lifesaving impact on such age-related diseases as stroke, hardening of the arteries, and memory loss
- its cancer-fighting properties
- its potential as a treatment for Alzheimer's disease, autism, and other conditions

In addition, we'll offer practical, low-cost strategies for incorporating these benefits into your life, starting today.

But first, consider the profound revolution in scientific thinking that these new discoveries represent.

Why Do We Age?

Throughout history, aging has been viewed by laymen and scientists alike as a straightforward and inevitable process: After years of use, the body just "wears out." Think of the body as a car: After 100,000 miles or so, parts and systems begin to fail, and eventually you just can't keep it on the road anymore.

That view seems to make sense. It's the basis for most of the methods doctors use to treat diseases of old age: repair, replace, or adjust. Fix the broken part if you can, patch it if you can't fix it, and remove it if you can't patch it. Implicit in this mechanical view of the body is the notion that aging and death are flaws in an otherwise elegant system, caused by inferior materials or shoddy workmanship.

The only problem is, we've known for at least 100 years that the analogy doesn't hold water. Bodies are not machines. They don't just "wear out." In fact, our bodies are renewing themselves continuously, replacing old cells with newly minted ones every day. (The only cells that don't get replaced are brain cells, those in the surface of the teeth, and, in women, the eggs within the ovaries.) Years of research at the most basic cellular and chemical levels suggest a different view altogether: Aging isn't caused by "flaws" in the body's systems. Aging *is* the system.

New Insights into Aging

We are also learning that aging is neither simple nor inevitable. Consider the following:

Many species of life are virtually immortal. Unlike the cells of the human body, single-celled organisms such as amoebas and bacteria never "die" in the traditional sense. Unless they are destroyed—say by heat, radiation, or by being eaten—they essentially live forever, reproducing by cloning genetic copies of themselves.

In more complex creatures, however, this system gets a curious twist. Human body cells, for example, reproduce in exactly the same way as free-living cells, but there's a limit on how many times they can reproduce before they simply die out. Cells taken from the body and cultured in the laboratory, we've found, can divide only about fifty times before they die (a phenomenon known as Hayflick's limit, after the biologist who first identified it). Cells taken from older people hit the limit even sooner.

This metabolic "time bomb" is found in complex, multicellular species, not in simpler and presumably older forms. That suggests that the aging process *evolved* as organisms grew more complex. (Interestingly, the only human cells that aren't subject to Hayflick's limit—the only ones that can grow and reproduce indefinitely in laboratory dishes—are cancer cells.)

Even among more advanced species, life spans vary dramatically. In California, a pine tree named "Methuselah" is 4,700 years old (and it's still sexually active!)—a fifty-fold increase over the average human life span. White mice—whose genetic makeup is so similar to humans' that they're used to test new drugs and therapies—have a life span of only two or three years—about one-thirtieth that of humans. Chimpanzees—our closest living animal relatives—have an average life span less than half that of humans.

Within species, however, there's surprisingly little variation in the aging process. If our bodies were *really* like cars, we'd expect to see at least a few true antiques on the road. And yet the oldest reliably reported life span is

120 years—about one and a half times the average human life expectancy. (In automotive terms, that would mean that the oldest car still running would date from the mid-1970s.) Think of those carnival barkers who make their living by guessing people's ages; they can win only because the physical signs of aging happen on a very tightly controlled schedule. Even those of us who are amateurs are able to guess most people's ages within five years or so, and be right nearly every time.

All of these observations suggest that aging isn't something that happens to our bodies as a result of the "slings and arrows" of everyday misfortune. Rather, it suggests that aging is controlled by an *internal* biological clock. And yet most strategies for treating and preventing the diseases of aging focus on controlling *external* factors, or on treating the symptoms of aging, not its causes.

As we mentioned above, the evidence suggests that the aging process is something that has evolved—just as, say, the ability to live on dry land evolved. On the face of it, that doesn't seem to make much sense. Why would old age be an evolutionary advantage? What possible benefit could it offer?

From an individual's perspective, not much. But from the perspective of an entire species, aging and death do indeed advance the cause of evolution by reducing competition between one generation and the next for food, water, shelter, and other scarce resources. Some years ago, then Colorado governor Richard Lamm created a national uproar by suggesting that older people have a responsibility to "die and get out of the way" of the next generation. That may have been bad politics, but it's good science. From an evolutionary standpoint, one generation's job is done when the next one reaches sexual maturity. Some researchers have advanced the same arguments in less colorful terms. As they point out, in a stable population the birth rate and death rate must cancel out each other. In short, the faster old generations die, the faster new ones can be born—and, therefore, the faster evolution can proceed.

In this sense humans are considerably less evolved than our more short-lived counterparts. Insects, for example, may live, grow, reproduce, and die in a matter of weeks or even days. This rapid generational turnover makes them ideal for studying evolutionary changes in the lab. It also gives them an edge in the real world: Some fifty years after the introduction of organic pesticides, many of the insect species they are supposed to kill have evolved into resistant forms.

Adjusting the Aging Clock

Despite the evolutionary advantages of death and aging, we can be forgiven if we do not wish to do our part for natural selection. Fortunately, if we understand how the system works, we can begin to change it. There is growing evidence that the aging clock can be reset.

Just as some people arrive at puberty early and some late, some also arrive at old age early and late. We all know individuals who just naturally seem to age more slowly than others, and for no discernible reason. (For every centenarian who credits his or her longevity to clean living, it seems there's another who attributes it to hard liquor and fried foods.) Beyond this normal variation, doctors also know that aging is affected by a variety of diseases and conditions. People with Down's syndrome, for example, have an average life expectancy of only about thirty-five years. They die young because they get older faster; in their late twenties they begin to show such typical signs of old age as baldness, dementia, and arthritis. Conversely, people with certain forms of blindness tend to live longer than usual, and age less quickly. Near-starvation also helps you live longer: In animal studies, scientists have found that they can very nearly double the average life span of mice by severely restricting their diet.

Over the past decade or so, researchers have discovered that the pineal gland and melatonin are the common denominator in these and other seemingly unrelated phenomena. *It now appears that the pineal gland is, in fact, the biological clock that controls aging, and that this gland's daily production of melatonin establishes the basic rhythm of life for every organ, every tissue, every* cell *within the body.* Melatonin also regulates other biological rhythms—the cycle of sleep and wakefulness and the onset of puberty, for example.

Laboratory studies are now buttressing these conclusions with evidence of direct links between melatonin and the aging process. For example, melatonin has been shown to protect us from many of the physical changes we associate with aging—everything from heart disease and menopause to memory loss and insomnia.

Light, Dark, and Ponce de León

Melatonin production is regulated, in large part, by the daily cycle of lightness and darkness (known as the *circadian rhythm*). The pineal gland is connected to the eyes, and darkness serves as a signal to the body to begin producing melatonin. When you feel sleepy in the evening, it's because the pineal gland has started pumping melatonin into your bloodstream, where it begins the physiological changes that prepare us for sleep: Heart rate and digestion slow, body temperature and blood pressure drop, alertness diminishes. In the morning, bright light hitting the retinas of the eyes shuts down melatonin production almost completely. These chemical

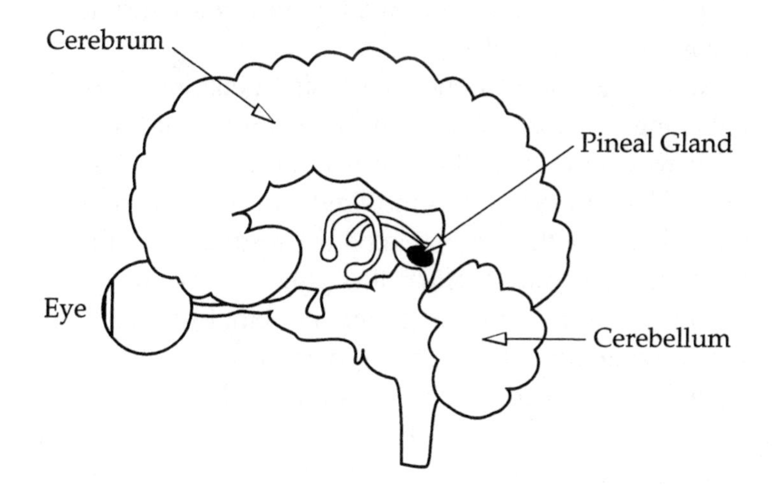

The Pineal Gland. *Located deep within the brain and connected by a direct pathway to the eyes, the pineal gland receives information about light and dark and translates it into a chemical signal that regulates rhythms throughout the body.*

changes promote sleep at night and keep us alert in the daytime. And as we'll see in later chapters, this basic cycle of melatonin production also helps protect the body from the effects of aging.

Our newfound understanding of this cycle raises some interesting possibilities about Ponce de León's quest and the annual migration of elders to the southern climes of Florida, Arizona, and California: Consider the fact that humans, who evolved in the tropics, aren't really well adapted to the seasonal swings in day length that we experience in the north. This balance is upset as the days grow shorter and gloomier, and as we spend more time indoors under artificial light and less time outside exposed to natural sunlight. It is only speculation, but it is not unreasonable to think that those Florida trips are promoted, in some subtle way, by our bodies' instinctive craving for more light.

And if, as the evidence suggests, melatonin controls the aging process, it just might be possible that Ponce de León was on to something in his quest for the Fountain of Youth—but that it was the *light,* not the waters or the warm breezes, that were behind the restorative virtues of the land he named Florida. It is almost certainly a coincidence, but an intriguing one nonetheless, that Mrs. Carrie White, the oldest known person ever to live in the United States, died in 1991 at the age of 116 years at a nursing home in Palatka, Florida—some fifty miles from a small spring in St. Augustine that, according to legend and local tourist brochures, is the long-sought Fountain of Youth.

In the chapters that follow, we'll explore in detail this exciting new research into melatonin. Even more important, we'll show you how these leading-edge discoveries from research labs around the world can improve your life *today,* and how you can use them to enhance your well-being. We'll show you how you can take advantage of your body's own natural production of melatonin, how to stimulate that supply, and how to augment it with safe and low-cost supplements. The benefits of this program include both immediate improvements in your health and long-term enhancements of your body's ability to fight disease and fend off the ravages of time. But first, let's take a closer look at the pineal gland and the key role it plays in our lives and well-being.

CHAPTER 2

The Pineal Gland: The Brain's Appendix

The ancients knew of it. The Greeks named it. The Romans described it. The Hindus associated it with the Crown Chakra, one of the seven spiritual centers of the body.

But nobody really knew what it was for. Greek anatomists noted that it was shaped like a minuscule pine cone—hence the name *pineal* gland. The Romans thought it might have something to do with the blood. Rediscovered by Renaissance anatomists, the pineal was thought to have a variety of ill-defined and unimportant body functions.

As late as the 1960s, if you'd asked leading physicians about the pineal gland, most would likely describe it as something like the appendix—a small, withered organ that might have served some long-forgotten purpose among our ancestors, but was now just an extra body part with no known function. As Russel J. Reiter, a leading pineal researcher from the University of Texas, puts it, "the pineal gland was until recently a marginal and beleaguered member of the neuroendocrine system."

Slowly, however, that view began to change. As early as 1898, reports had appeared in the medical literature suggesting that the pineal gland had some effect on puberty. But these reports were confusing and contradictory. In one, for example, a young boy who developed a tumor in the region of the pineal gland experienced puberty shortly thereafter. But in another report, a tumor in a similar location apparently *delayed* puberty. (Though the reason wasn't known at the time, these contradictory events probably occurred because the first tumor stimulated the pineal gland while the second one attacked it.) It was known that in birds, the pineal gland was much larger and robust, and that it was connected by nerves to the eyes. But still nobody knew its function, in birds or in humans.

In the late 1950s, the mysteries of the pineal finally began to emerge, though it would take another three decades before the pieces of the puzzle began to form a complete picture. It began when researchers learned that the pineal gland secretes a substance found to be chemically related to the hormone serotonin and to the skin pigment melanin; thus, it was given the name *melatonin*.

But for the next decade, nobody knew what melatonin did. The pineal gland produced it in such minute quantities that even the most sophisticated laboratory instruments could barely detect it in the bloodstream. In 1963, scientists learned that melatonin was in fact a hormone—one of the chemical messengers that helps the body regulate a variety of functions. But stymied by the limits of their instruments and their knowledge, they still weren't able to discover the function of this hormone.

In the early 1970s, Dr. Reiter and others began to unlock the secrets of the pineal gland and its mysterious product, melatonin. Researchers had been hampered in their efforts by the scarcity of melatonin, which the pineal produced in virtually undetectable amounts. Eventually the researchers learned how to make synthetic melatonin in relatively large quantities, and they found easier ways to measure the microscopic quantities that circulated in the blood. The synthetic material was chemically identical to natural melatonin, and its low cost and availability made it possible to conduct far more extensive experiments.

As the studies began to accumulate, it became apparent that the pineal gland was actually a major-league organ, with far-reaching effects:

- In the pituitary gland, melatonin acts as a "master hormone," stimulating the release of a wide variety of other hormones. These hormones, in turn, regulate numerous bodily processes, ranging from digestion to menstruation.

- In the brain, melatonin excreted by the pineal gland acts as a sleep-inducer, depressing brain-wave activity and preparing us for sleep.

- In the heart and circulatory system, melatonin reduces the likelihood of blood clots forming, which in turn helps protect us from heart attacks and stroke.

- In the bloodstream, melatonin enhances the ability

of white blood cells to form antibodies—the chemical signals that the body uses to identify and fight off infections.

- Throughout the body, melatonin acts directly on cells as an antioxidant, protecting them from damage from free radicals—chemical compounds that have been implicated in cancer and many other diseases.

And, as we'll see in subsequent chapters, these findings are only part of the story. In fact, wherever researchers have looked for melatonin, they've found it—performing a vast array of critical life functions.

Inside the Network

How can a chemical produced in such small quantities have such big effects? And how can it possibly affect so many different parts of the body? Part of the answer lies in melatonin's role in helping the body talk to itself.

The human body is made up of a vast and various collection of organs and tissues performing myriad different functions. To keep all of them working in harmony requires a communication system that makes Ma Bell seem like a little old lady.

Information is transmitted throughout the body in two ways: electrically and chemically. Sound entering our ears, for example, is transformed into tiny electrical pulses that travel into the brain via the auditory nerve. Signals from our eyes travel in much the same way via the optic nerve and other pathways. Electrical signals from the brain, in turn, control the muscles—not only the voluntary muscles that we use to walk and swing a hammer, but the involuntary muscles of the heart, lungs, digestive system, and other functions.

But this is only part of the network. In addition to these electrical signals, there are *chemical* signals. One group of chemical messengers, known as *neurotransmitters,* helps carry signals across the synapse, or space, between nerve cells.

Neurotransmitters are believed to play a role in virtually every function within the brain—sleep, alertness, pain, pleasure, emotions, anxiety, memory, impulse, moods. In a sense, they help assign meaning to the raw information generated by our senses and thoughts. In fact, many of the "mood disorders" such as depression and mania show up as imbalances in the brain's neurotransmitters, and many of the drugs used to treat these disorders do so by influencing these chemical messengers.

Hormones—including melatonin—are closely related to neurotransmitters. Produced primarily in the various glands of the *endocrine system,* they help regulate how cells function throughout the body. Often they have wide-ranging effects, coordinating the activities of numerous systems. For example, adrenaline—the hormone produced by the adrenal glands—triggers a range of changes in response to stress: It speeds up the heart rate, shifts the blood supply to essential organs, slows digestion, and heightens alertness.

There's a lot of overlap between these two groups of chemical messengers. Serotonin and dopamine, for example, function as both neurotransmitters and hormones, depending upon where they are found in the body.

The pineal gland has a foot in both camps. Biologically, it is composed of nerve cells, which are closely related to the nerve cells found in the retina of the eye. Functionally, however, the pineal is considered part of the endocrine system, which consists of the pituitary, thyroid, thymus, parathyroid, and adrenal glands, as well as the reproductive organs and the pancreas. Each has multiple functions.

The endocrine system isn't a single physical entity; it's a rather loose association of organs with various functions. To qualify for membership to the system, the organ, called a gland, must secrete a substance within the body (that's what *endocrine* literally means).

For most of medical history, the glands of the endocrine system were pretty mysterious. Science lacked the ability to isolate, measure, and identify the minute fluids that they secreted. Since the middle of the twentieth century, however, the field of endocrinology has been revolutionized, as more sophisticated laboratory methods have helped scientists study these secretions and their effects. At the same time, advances in biochemistry have allowed us to manufacture synthetic versions of these hormones in large quantities, making it easier to study their effects and to use them for medicinal purposes. Doctors use them in much the same way our bodies use the natural substance—to regulate and modify basic bodily functions. For example, corticosteroids, which are produced by the adrenal glands, affect the immune system (among many other functions); as a drug they're useful in reducing inflammation and allergic reactions.

The pineal gland, though tiny, occupies an important

role in this system. Like the conductor at the orchestra, it coordinates the actions of the other glands. This helps explain why melatonin affects so many different parts of the body.

The Melatonin Message

Melatonin plays a key role in how both the chemical and electrical "networks" interact within the body. Through a carefully orchestrated series of steps, the body translates information from the outside world into a chemical message that reaches every part of the body and helps keep this complex system in harmony. The message begins in the eyes, where light falling on the retinas produces a nerve impulse. The eyes are "wired" into the pineal gland by a nerve pathway. (In fact, evidence suggests that the pineal originally evolved from cells of the eye.)

When this impulse arrives at the pineal gland, it coordinates a series of chemical reactions resulting in the production of the hormones serotonin and melatonin. When the eyes detect light, the pineal gland produces serotonin and virtually no melatonin. When darkness comes, the pineal begins to convert serotonin into melatonin. Thus, working in concert, the eyes and pineal gland translate information from the outside world (light and darkness) into a chemical message (serotonin and melatonin) that can be decoded by every cell in the body.

The pineal gland doesn't store melatonin as it's produced; it pumps the hormone directly into the bloodstream. Thus, during the night relatively high levels of melatonin circulate through the bloodstream into every part of the body. When light from the eyes shuts down the pineal's production, the melatonin levels in the bloodstream and tissues fall almost immediately.

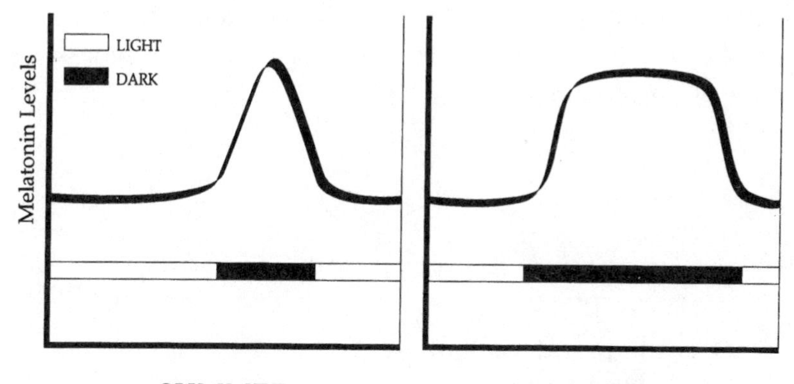

Melatonin Production. *At the onset of darkness, melatonin production begins to rise, peaking in the middle of the night and declining toward morning. As the period of darkness lengthens (during the winter), melatonin is produced over longer periods as well.*

Decoding the Message

The differences in melatonin levels throughout the day (low by day, high by night) help to coordinate the body's functions into a rhythmic, synchronized system. The chemical message that the pineal gland sends out is simple but profound: It tells the body what time it is.

The second part of the message comes from the

amount of melatonin that is circulated throughout the body. When we are young, our cells get a big shot of melatonin every night. As we grow older, these nightly peaks of melatonin production become lower and lower. It appears that this diminution is responsible for the aging of cells, and sets the stage for damage and eventual disruption of vital bodily processes—in other words, the physical decline of old age.

Toward a New Theory of Aging

Over the past decade, these new insights into melatonin have slowly begun to assemble, like pieces of a jigsaw puzzle, into a broader picture. As we'll see, melatonin is emerging as the key connection among several of the most basic life functions—the ability to reproduce, to fight off disease and infection, to regulate rhythms day to day and across the life span. All of these functions are critical to the body's ability to perform. And all of them, when they go awry, are manifestations of the process we call aging.

Aging isn't just a random process of disintegration and decline. On the contrary, predictable things happen when we get old. For example, we're more prone to certain diseases, such as heart disease, cancer, and diabetes. We're less able to fight off infections. We don't sleep as well at night. Memory fades. Women become infertile, and men lose some of their sexual potency.

It soon became apparent to a handful of researchers that all of these phenomena were, in one way or another, related. And as we'll see in the next chapter, these various lines of inquiry rapidly converged into a revolutionary new view of aging, with the tiny pineal gland—once dismissed as "the brain's appendix"—playing the starring role.

CHAPTER 3

Melatonin and Aging

A group of scientists gathered some years ago in the Italian town of Copanello to discuss the latest research on problems of aging. Amid the countless dense dissertations on biochemistry and endocrinology, on statistics and methodology, one presentation stood out for its elegance and real-world implications. Its provocative author argued that it was possible to dramatically extend the life span of mice in the laboratory by restricting their diet. This first early evidence that the aging process could be controlled represented a turning point in our understanding of biology. But on that day when it was presented to this conference of some of the world's foremost experts on aging, it was not met with applause, but with laughter.

Impossible, skeptics said. The experimenters must have been mistaken. The notion that all you had to do to slow the aging clock was to eat less was too good—too *simple*—to be true.

Pineal researcher Walter Pierpaoli, writing in the *Annals of the New York Academy of Sciences,* recalls that he first started believing in the existence of an aging clock when he heard the laughter in the audience. Quoting a colleague, he wrote that there were three ways a researcher knew whether his or her scientific discoveries were valid: "First, they were valid if other scientists said they were wrong; second, if others began laughing at him; and the final test of the validity of his discoveries came when he found that he was no longer receiving grants."

And when researchers later demonstrated that you could achieve the same kind of life extension by giving mice daily doses of a naturally occurring, low-cost hormone, the skepticism intensified.

For one thing, this new evidence flew in the face of some cherished notions about old age. The prevailing view of aging, emerging from decades of study and billions of dollars of research, has been that it is a complex process, involving numerous factors over which we have relatively little control. Consider, for example, the complexities of just one aspect of aging: the rising risk of cancer. There are scores of cancers and no two are exactly alike. Even within a single type of cancer, there's a bewildering array of variation. Most lung cancers are related to cigarette smoking, but some occur in people who've never smoked a cigarette in their lives. Meanwhile, many three-pack-a-day smokers *never* develop cancer. Some people get lung cancer in their forties, some in their eighties. There are different types of lung cancer—some more aggressive, some less so. Some respond to treatment; others don't.

And that's just one tiny piece of the whole aging

puzzle. Despite intensive research, nobody knows why Alzheimer's disease occurs, or how to prevent it. The incidence of heart disease rose rapidly in the first six decades of the twentieth century, and then, for no discernible reason, tapered off and actually began to decline.

Methusalean Mice

Even after scientists accepted the findings that mice could be kept alive longer by limiting their food intake, not much changed on the frontiers of geriatric medicine—at least not immediately. For one thing, the discovery had little practical value, since the mice had to be kept on the verge of starvation, starting in infancy. Most researchers into the problems of old age regarded the results as little more than a scientific curiosity. But they were the earliest indicators that we could indeed regulate aging.

Those studies laid the groundwork for inquiries into melatonin's antiaging effects—studies that were harder to ignore. Further research showed that underfeeding not only extends life span, but preserves the youthful rhythm of melatonin production, suggesting that this is how food restriction extends life span. These findings led researchers like Pierpaoli to see if it would be possible to achieve the same results by administering melatonin directly rather than by limiting dietary intake.

It was. Eventually, other scientists found similar results. But still questions remained. Though mice did indeed live longer than normal in these experiments, it was difficult to sort out the effects of melatonin and the pineal gland from other factors, such as food intake. Was the extended life span really caused by melatonin and the pineal gland?

Then, in 1991, Pierpaoli and a Russian scientist, Vladimir Lesnikov, reported a dramatic experiment that gave direct and incontrovertible proof of the pineal gland's control of the aging process. They bred two groups of mice that were genetically identical in every respect, and raised them under identical conditions. There was only one difference between the groups: The first group was young—three to four months of age; the second group, at an age of about eighteen months, was well into retirement age in mouse years.

After anesthetizing the mice, the researchers opened the rodents' skulls and *swapped the old and young pineal glands*. All of the "old" mice now had youthful pineals, while the young mice had aged glands.

At first, not much happened. Because all the mice were genetically identical, their bodies accepted the new organs without any complications. Then, as the weeks passed, the young mice began to fail, showing unmistakable signs of accelerated aging. The older mice were rejuvenated. By the end of the experiment, there was little doubt about the pineal gland's role in aging: The young mice with the "old" pineals lived an average of only seventeen months—about two-thirds of the normal life span. The old mice who'd received the "young" pineal glands, by contrast, lived for *thirty-four months—twice as long as the first group, and almost half again longer than the normal life span!*

New View of Old Age

Out of these studies has come a new and different view of aging, one that regards it not as a collection of unrelated problems but as a progressive deterioration of basic body systems.

Scientists now look at aging processes in a number of ways:

- As a breakdown of *bodily rhythms,* in which the carefully orchestrated systems of the body begin to fall out of sync. According to this view, for example, memory losses associated with aging may be related to disruptions in sleep patterns.

- As a breakdown of *communications,* in which the elaborate communications system we described in chapter 2 begins to run less efficiently. As communications worsen among various systems, the efficiency of the body as a whole declines. According to this theory, for example, the endocrine system of glands becomes less sensitive to, say, changes in the cardiovascular system, and so produces the wrong amounts of hormones, or releases them at the wrong times. Alternatively, the body's cells become less sensitive to these hormones and can't interpret the chemical "message" they carry.

- As a breakdown of the *immune response* and the body's ability to recognize differences between itself and foreigners. On the one hand, this diminished capacity leaves us more susceptible to infections and cancer and slower to heal; on the other hand, it leads to an increased incidence of autoimmune disorders, which occur when the immune system fails to recognize the body's own tissues and attacks them as if they were germs.

There's a lot of overlap among these three views, and to a certain extent they're simply three different perspectives of a bigger picture: the loss of stability in body systems as a whole. For example, communications breakdowns will naturally lead to disruptions of body rhythms, which in turn will impair the ability of the immune system to coordinate a response to infections and cancers.

These three views all come together at the pineal gland. They reflect the three key roles of the pineal and its most important product, melatonin:

- As an internal clock and calendar, melatonin coordinates the rhythms of the body.
- As a chemical messenger (neurotransmitter and hormone), melatonin helps maintain an efficient flow of communication.
- As a master hormone, melatonin plays a key role in regulating the immune system.

Melatonin and the Digestive System

In addition to the pineal gland's rhythmic production of melatonin, there's a second major source of melatonin in the body: the digestive tract. Intestinal melatonin, unlike that of the pineal gland, is produced relatively constantly. It establishes a "baseline" level of melatonin throughout the day. What's more, melatonin production in the gut isn't affected by light and dark. In short, the digestive system appears to be an entirely independent source of melatonin.

Researchers are just beginning to explore the relationship between the gut and melatonin, and it's not yet clear why this second source exists, or the role it plays in aging. Significantly, however, they've learned that melatonin production increases in the digestive tract when calories are restricted. Some researchers speculate that this effect may be part of a complex adaptation of the aging process to respond to famine conditions. When food is scarce, fertility declines. And those young animals that *are* born stand much less chance of reaching sexual maturity: Being younger and smaller, they're at a disadvantage in the competition for scarce resources.

With fewer offspring surviving, the survival of the community may well depend on extending the fertility and life span of the older generation. In simple terms, the older generation needs to live longer so that it will have more opportunities to "replace" itself. When food is plentiful, by contrast, more offspring survive and reproduce. Under such conditions, the aging process accelerates and individuals die sooner to make room for the new arrivals.

If this theory is correct, the sensitivity of melatonin to food supplies is critical to help fine-tune fertility and survival to available resources. Just as the eyes gauge light and dark, regulating the pineal's production of melatonin, so the digestive system gauges food supplies, adjusting melatonin levels in response.

In addition, this theory may offer some insights into

why "lifestyle" diseases such as heart disease and cancer so often seem to be associated with rising prosperity. By removing the threat of famine, modern industrialized societies may be unwittingly sending a signal to our bodies to accelerate the aging process.

Hormones and Aging

From studies on the pineal gland we've learned that aging isn't something that afflicts our hearts, our kidneys, our skin, our minds. The changes that occur in these and other organs are the outward *symptoms* of a process that begins at a hormonal level.

And aging isn't something that just happens to us when we're old. It starts even before we're born. At different stages of life, we call it by different names. In infants and small children, we call it childhood development, and it's generally viewed as a positive process: As children's bodies mature, they acquire important abilities and skills. In later years, we call the process puberty and then adolescence, as the body undergoes physical and mental changes. Then come two or three decades in which the body becomes less efficient. We may put on weight, lose our hair; our skin becomes less elastic and more prone to damage; we can't reach those long fly balls anymore. Health problems begin to come up more frequently. Even so, the physical changes between the ages of, say, nineteen and forty-five occur far more gradually than those that come before and after.

As we move toward our sixth and seventh decades, change comes more quickly once again. In women, menopause produces profound physical alterations that go beyond the question of fertility. In men and women alike, significant and potentially fatal health effects begin to occur: heart disease, hypertension, high cholesterol, increased risks of cancer, changes in brain function, and others. For different people, the pace is different but the patterns are similar; ultimately one or some combination of these degenerative diseases leads to death.

How We Age

At each of these stages of life, the pineal gland undergoes corresponding changes. In the first three months of life, the pineal gland secretes little or no melatonin. Once melatonin production begins, however, it really picks up steam. As children, our melatonin levels are at the highest they'll ever be in our lives.

The changes of puberty are now believed to be triggered by declining melatonin levels in the blood. Interestingly, though, melatonin *production* remains fairly steady up to and through puberty. What changes is body size: As we grow larger, the *concentration* of melatonin in the bloodstream becomes less; in effect, the same amount of melatonin must now serve a larger body. (This finding may shed some light on the link between obesity and such disorders as heart disease. As the body gets bigger, the concentration of melatonin may become diluted, giving rise to these degenerative diseases; see Chapter 16.)

After adolescence, however, the pineal gland gradually begins to reduce the amount of melatonin that it produces for a number of reasons: Like other parts of the brain, the pineal gland doesn't replace lost cells; once they're gone, they're gone forever. Also, calcium deposits build up within the pineal gland over time. It's not clear what effect this calcification has, but it seems to interfere with the gland's ability to function efficiently. One group of researchers speculates that the accumulation of calcium deposits in the pineal gland gradually diminishes the function of the pineal gland. In old age, this degeneration reaches the point that the pineal produces virtually no melatonin whatsoever.

A Unified Theory

This research on the parallel progression of aging and decrease in melatonin levels, along with the laboratory experiments, now gives us a revolutionary new view of the aging process. It isn't just a collection of unrelated problems that overtake us as the years go by. It's a carefully orchestrated process that begins in the pineal gland. And most significant of all, it's a process that can be *controlled*.[1]

In the chapters that follow, we'll look at the many practical implications of this new view of aging, and at the ways in which you can put melatonin to work for you today. But first, consider the profound shift in scientific thinking that this new view represents.

Of course, we have not reached the point that we entirely understand the mysteries of aging—and we probably never will. But now we have a new way of looking at the problems of aging—a view that gives us insights into how many of the seemingly unrelated aspects of aging may in fact be interrelated at the most fundamental levels of biology. Melatonin research is giving us, for the first time, a unified theory of aging.

This new view of aging can lead to new insights and encourage doctors to "look over the fence" of their specialties in the search for better treatments. For example, we have known for some time that treatment with bright lights is useful for some kinds of depression. Knowing that this effect occurs by way of the pineal gland, and knowing that the pineal gland also influences the immune system, researchers may consider looking into the effects of light on our resistance to infection.

This kind of fundamental shift in thinking occurs rarely in medicine. When Louis Pasteur demonstrated that germs cause many kinds of disease, he launched a line of inquiry that ultimately resulted in antibiotics to treat a vast range of conditions. Similarly, when researchers in the 1950s found that certain kinds of drugs could relieve psychiatric symptoms, it dispelled some of the most deeply held myths of mental illness. It helped doctors understand the chemical basis of emotions, moods, even sanity itself—forever altering both the way we treat and the way we think of psychiatric disorders.

Pineal research is now having the same kind of fundamental impact on our notions of aging and the diseases that are related to it. Aging is seen less and less in metaphysical terms—as a process that is largely beyond the reach of medicine—and more and more in chemical and physiological terms, as a process that can be influenced by medical intervention, as a *treatable* condition.

These new discoveries will not, of course, close down the heart centers, nursing homes, or oncology clinics, but may push the diseases they treat off to later years and help prevent many cases of cancer, heart failure, and other diseases of aging. Nor will an increase in melatonin replace the need for exercise, a low-fat diet, and other elements of a healthy lifestyle, but, as we'll see in the second part of this book, it may add to their benefits. In short, this theory of aging gives us new ways of looking at one of mankind's oldest mysteries—and practical strategies that we can begin to follow now.

Later in the book, we'll look at how you can implement those strategies in your own life. In the chapters immediately following, we'll explore the specifics of how melatonin affects health problems and diseases throughout the life span, and the promise it holds as a new approach to treatment for a vast array of medical conditions.

CHAPTER 4

Melatonin and Free Radicals

Central to melatonin's role as an antiaging hormone is its action as a free-radical scavenger. Free radicals can wreak havoc on cells, and many scientists believe that free-radical oxidation is the fundamental mechanism of aging.

Free radicals are incomplete atoms and molecules. In ordinary chemical compounds, electrons orbit in stable patterns. Free radicals, however, are short one or more electrons, making them unstable and prone to combine with other compounds and disrupt their structure. In living cells, these renegade compounds upset the complex and delicate chemical structures of life itself. In sufficient numbers, they can break up components of the cell, killing it outright.

Many scientists also believe that even when free radicals don't destroy the cell completely, they can inflict permanent damage. According to this free-radical theory, this damage underlies the deterioration of body cells and tissues that we see in aging. For example, wrinkled skin is really the result of the skin's collagen structure breaking down. When we're young, this structure is pliant and supple, and it supports the skin against the downward forces of gravity. The main factor in triggering its breakdown is exposure to ultraviolet rays, primarily from sun exposure, a known and potent generator of free radicals. (The pigment in the skin, melanin, is chemically related to melatonin, and it protects against this ultraviolet attack. That's why darker-skinned people often don't wrinkle as quickly as fair-skinned ones.)

Similarly, free-radical assaults on the hair follicles may, over time, disrupt their functioning, causing them to lose their ability to produce pigmentation. As a result, the follicle begins to produce hair that is without color, or gray.

Throughout the body, the free-radical theory of aging contends, these and similar kinds of breakdowns contribute to the physical signs and symptoms of aging. Muscles become less powerful, broken bones knit more slowly, the kidneys become less efficient, memory becomes more erratic.

If free radicals reach the DNA within the cell's nucleus, they can also cause more subtle but, in the long run, more devastating damage. By breaking the delicate bonds in DNA, they can alter the genetic code and cause the cell to mutate. In many cases these genetic mutations are lethal to the cell; in others, however, they can make the cell become cancerous. As we learn more about cancer, it seems that this kind of genetic damage is at the root of many—perhaps most—types of cancer. Tobacco smoke, for example, generates high levels of free radicals in and around the lung's cells. Some people's heredity makes them more susceptible to genetic damage—that's why, for example, cancers run in families—but in most cases environmental conditions probably act as the catalyst that triggers the development of the cancer cells.

Some researchers believe that melatonin's first function was to protect early life from such damage. In fact, that may explain why melatonin is produced in daily rhythms: In early organisms, the greatest risk of damage would occur at certain times of the day, when light and radiation from the sun interacted with chemicals in the "primordial soup" of ancient oceans. According to this theory, the daily fluctuation of melatonin that occurs in virtually all animals started as a reaction to the highs and lows of free-radical levels throughout the day. Later, as animals became more complex they took advantage of this rhythm, using it as a built-in "clock" to regulate various body rhythms and synchronize them to the sun.

Ancient Protector: One Theory of the Evolution of Melatonin

We suspect that melatonin must have evolved very early in the history of life on earth, because it's present, in the same chemical form, in virtually every cell where researchers have looked for it. Some researchers believe that it evolved to counter the world's first instance of air pollution: The release of oxygen by green plants.

Today, of course, we need oxygen to live, but for early life it was probably life-threatening. Oxygen is extremely reactive with other chemicals—that's why iron rusts, paint oxidizes, and coal burns. Early life evolved in a low-oxygen environment. When plant life developed the process of photosynthesis—converting sunlight into chemical energy—it began pumping vast quantities of oxygen into the air as a by-product. As oxygen levels rose from virtually zero to approximately 21 percent of today's atmosphere, early life forms had to find a way to protect themselves from it.

A theory advanced by researcher Russel Reiter suggests that melatonin evolved first as a protection against damage from rising levels of atmospheric oxygen. That would explain its rhythmic production. Free radicals tend to be produced during the day—caused by the interaction of sunlight and oxygen—rising to peak levels toward evening, when melatonin levels begin to increase. By morning, many of these free radicals have broken down, and their concentration is at its lowest point during the day.

As organisms grew more complex, this theory suggests, this basic daily rhythm of melatonin production was used to regulate a variety of daily rhythms.

Free radicals are still with us. In modern humans, they're implicated in a wide variety of diseases besides cancer, including Parkinson's disease, multiple sclerosis, muscular dystrophy, rheumatoid arthritis, emphysema, circulatory disorders, and cataracts. All day long, we absorb them into our bodies from the air we breathe, the food we eat, the water we drink, the chemicals we come into contact with (see table on page 43). In addition, our bodies produce free radicals themselves—for example, to kill foreign invaders. (That's why infections are a source of free radicals.) Just as in the ancient days, free radicals are produced more by day than by night, meaning that their levels start out low in the morning and rise during the day, peaking in early evening.

Within the cell, melatonin acts as an efficient "scavenger" of these free radicals. Whenever it encounters free radicals, melatonin chemically binds with them. The resulting compounds are inert and unable to form the unstable chemical reactions that damage cellular structures. Now harmless, they're excreted from the cell and, ultimately, the body.

Clearly, the more melatonin that's circulating, the more powerful will be its effects on free radicals. As levels diminish, more free radicals are left behind to cause damage. According to the free radical theory of aging, as melatonin levels decline with age, fewer free radicals are neutralized. It's sort of like what happens if you don't put enough detergent in the washer: Some of the dirt is washed out, but some is left behind.

Free Radicals, Cholesterol, and Heart Disease

It now appears that free radicals play a key role in the formation of arterial plaque—the thick deposits that form on the inner surfaces of arteries and block the flow of blood. Atherosclerosis, as this condition is known to physicians, is the leading cause of heart attacks: As plaque blocks the narrow, winding blood vessels that snake along the surface of the heart, it cuts off the supply of oxygen-rich blood to the heart muscle. Such blockages are also the most common cause of strokes, in which blood supply to parts of the brain is impaired.

Sources of Free Radicals (a partial list)

Oxygen (breathing)	X rays
Cigarettes	Some medications (certain anti-cancer drugs)
Alcohol	Foods
Ozone	Exercise
Ultraviolet radiation	Bodily processes (metabolism)
Pesticides	
Carbon monoxide	

In bypass surgery, surgeons build "detours" around these blockages. Treatments such as balloon angioplasty open up passages through the blockage and restore the flow of blood. But neither of these treatments gets at the underlying cause. In fact, the plaque continues to accumulate, and the reopened blood vessels often close up again—sometimes after many years, sometimes within a few months.

Scientists have long known that cholesterol—specifically, the "bad" type of cholesterol known as LDL cholesterol—makes up a large component of the plaque material. (Another type of cholesterol, HDL cholesterol, doesn't contribute to the problem; in fact, it seems to offer some health benefits.) Now it appears that free radicals may act on the cholesterol, oxidizing it in a process not all that different from what happens when cooking oil goes rancid or milk curdles.

Many researchers believe that when infection-fighting cells in the arterial walls encounter these rancid particles of cholesterol, they attack and absorb them. Unfortunately, the cells can't break down the cholesterol and it accumulates. Eventually, the cells become thick and bloated, and as more and more of them swell up, they begin to block the vessel. As circulating blood slows around the obstruction, clotting factors in the blood are also activated, and clots form around the obstruction as well. Ultimately, all of these changes contribute to the formation of the thick, gooey plaque material.

Researchers are just beginning to explore the interaction of free radicals and cholesterol in the formation of arterial plaque, and they haven't yet established definitive answers. But numerous studies have already linked diets high in antioxidants with a reduced risk of heart disease. And if the theory is correct, it suggests two independent ways to reduce the formation of arterial plaque: by reducing cholesterol and by reducing free radicals. In other words, melatonin and other antioxidants may confer some protection from the effects of high cholesterol.

Antioxidants

In the past several decades, scientists and laypersons alike have heard a lot about antioxidants and free-radical scavengers. (The terms are interchangeable; the damage caused by free radicals is called oxidation.) A variety of substances, ranging from the beta-carotene found in tomatoes and carrots to vitamin E, have antioxidant properties and have been touted as protection against free-radical damage.

Numerous studies demonstrate the ability of antioxidants to head off trouble, especially for those disorders that are associated with aging. For example, a five-year study in China, reported in the *Journal of the National Cancer Institute,* found a 13 percent reduction in cancer deaths and a 10 percent reduction in death from strokes among people who took a combination of antioxidant supplements, including beta-carotene, vitamin E, and selenium, versus those who were taking a placebo.

Another age-related disorder, cataracts, also appears to be the result of free-radical damage, most likely caused by ultraviolet (UV) light entering the eye. That's why sunglasses with UV-blocking coatings help prevent cataracts. Studies by the Department of Agriculture show that people whose diets are high in antioxidants are six times less likely to develop cataracts. In addition, cataracts are more common among people with low blood levels of vitamin C and beta-carotene. A British study found similar results: Women whose diets were high in beta-carotene had a 40 percent lower risk of developing cataracts. A U.S. study found similar effects for vitamin E: It appeared to protect not only against cataracts but also against macular degeneration, a progressive disease that attacks the retinas and can cause blindness.

Scientists have found that antioxidants also have beneficial effects on the immune system. High doses of vitamin E—800 international units (I.U.)—improved the immune systems of older adults in one experiment. The current Recommended Daily Allowance (RDA) for vitamin E is only 10 I.U., and this study raises the question of whether that figure should be revised upward. In my practice, I've found optimal doses of vitamin E to be 400 to 600 I.U. Beta-carotene supplements of 180 milligrams per day have been shown to increase the number of T-helper cells—cells that help the immune system respond to attack. Daily supplements of 1,000 milligrams or more of vitamin C also have been shown to enhance the immune response. And according to the medical journal *Lancet,* vitamin E, zinc, and selenium all have been shown to augment the immune system as well.

Studies also show that people whose diets are rich in vitamin C tend to have lower blood pressure and healthier cholesterol levels. Similarly, foods high in beta-carotene also help protect against heart disease. In a study at Harvard University, physicians who suffered from heart disease took 50 milligrams of beta-carotene a day (about the same as you'd get from two cups of cooked carrots). Compared with a control group that didn't take beta-carotene, these doctors had only about half as many heart attacks, strokes, and heart-related deaths. Other Harvard studies revealed that women and men who take vitamin E also have a reduced risk of heart disease.

In addition to such well-known antioxidants as vitamins C and E, you can expect to hear more about CoQ10, flavinoids, and glutathione. They've been known to nutritionists for years, and now they're getting broader attention for their antioxidant and antiaging properties.

Melatonin: The Most Potent Antioxidant

Among antioxidants, melatonin is unique for a couple of reasons. First, it is the *most efficient* free-radical scavenger ever discovered, and it's especially effective against the "–OH" radicals—those that contain an oxygen and hydrogen compound that makes them especially active. Second, melatonin is by all evidence completely harmless to the body. No matter how high the levels, it apparently causes *no* side effects other than drowsiness. And unlike other antioxidants, which can become chemically unstable once they combine with free radicals, melatonin remains stable. As other antioxidants break down and release the radicals back into the environment, this process can actually *accelerate* cellular damage under some conditions.

Within the cell, melatonin gives special protection to the nucleus—the central structure that contains the DNA. By protecting DNA, melatonin protects the integrity of the cell's "blueprint." A cell with structural damage but intact DNA can usually repair itself easily, but a cell with damaged DNA often cannot fix even minor damage. This affinity for the nucleus suggests a particular ability to protect against chromosomal damage that can result in cancer.

Free Radicals and Old Age

Melatonin's fight against free-radical damage is central to its antiaging properties. The general cellular damage inflicted by free radicals is the driving force behind the aging process. Without melatonin's protection, cells would succumb quickly to this onslaught, giving rise to a progressive loss of function and rhythmicity.

All organisms generate free radicals as a by-product of metabolism. All of them have developed ways to minimize the quantities they produce. And all of them have a variety of defenses to neutralize them. The biological evidence is clear: Species with the most vigorous, successful defenses against free radicals are the ones that live longest.

As we've seen, the body's cells become more susceptible to damage as melatonin levels decline in old age. But free radicals aren't equal-opportunity ravagers; some systems get hit harder than others. In fact, this is some of the strongest evidence in support of the free-radical theory of aging: *Those body systems where free radicals are most common are the ones that deteriorate the most as we age.*

For example, memory loss is closely associated with production of the amino acid glutamate in the brain. On the whole, glutamate is a "good" chemical, though it has some nasty habits. It helps sculpt and shape the networks of neurons within our brains that we use to organize thoughts, actions, and memories. It establishes chemical pathways among these neurons that help guide the signals they receive. By reinforcing these pathways, glutamate helps build the patterns of our very consciousness.

But these networks tend to be hard on their component nerve cells. Like an interstate highway, the neuronal networks get a lot of traffic—and with it, a lot of wear and tear. The constant firing of synapses releases free radicals; over time, they kill the nerves along the network (which is a compelling argument against rigid thinking!).

Glutamate also causes damage by generating free radicals directly. In the laboratory, the brains of animals exposed to high levels of glutamate begin to show signs of premature aging—not only memory loss, but behavioral changes and nerve cell degeneration.

Melatonin fights these free radicals in the brain. It also

protects the nerve cells by keeping the neuronal networks from becoming too fixed. It promotes the development of alternate pathways, thereby helping to prevent overstimulation of the neurons.

But as we age and melatonin levels decline, the body count along the neuronal highway starts to climb. This loss of brain cells results in the confusion and impaired memory associated with old age. And significantly, research shows that the *greatest cell losses occur in precisely those parts of the brain that are most affected by glutamate.*

Similar patterns occur throughout the body, linking the processes of aging with free-radical damage. And everywhere that free radicals are found, it's a good bet that melatonin is on site as well, protecting the body's systems from the ravages of chemical attack.

CHAPTER 5

Melatonin and the Immune System

In light of melatonin's ancient role as protector, it is perhaps not surprising that we should find it playing a key part in the body's modern defense system. In fact, the pineal gland is—figuratively and literally—at the very top of the immune system.

The body's ability to fight off harmful invaders requires a lot of orchestration and communication. First the body must recognize danger; then it must quickly rally and organize its defenses. At the same time, the body alters its most basic functions to help repel the invader. Some suggest, for example, that fever may be the body's way of creating less hospitable conditions for germs and that runny noses and chest congestion help trap and expel them from the body.

Just as modern warfare is as much about logistics as individual bravery, so has melatonin's role evolved from roving warrior, protecting cells from free-radical damage, to a commander in chief overseeing a full-scale campaign. To be sure, melatonin continues to act on behalf of individual cells, but it also helps regulate the complex components of the immune system. This system is controlled in large part by hormones, most of which, in turn, are regulated to greater or lesser degrees by melatonin. As a "master hormone," melatonin helps orchestrate the many and varied functions of the immune system, helping them work at peak efficiency.

As any general will tell you, though, an effective fighting force requires more than bravery; it requires discipline. And while melatonin helps marshal a vigorous defense of the body, it also helps keep the immune system from becoming *too* aggressive—running out of control and wantonly attacking the body's own tissues. When this discipline breaks down (as it is more likely to do as we get older), we become prey to *autoimmune* disorders, such as diabetes and lupus, in which the immune system attacks the body's own cells.

What Is Immunity?

To understand the role of melatonin in protecting us from disease, let's take a closer look at the immune system.

First, the word "system" is something of a misnomer. It's actually several systems that work in concert to provide multiple lines of defense. These defenses are grouped into two broad categories: humoral immune mechanisms and cell-mediated mechanisms.

Humoral Mechanisms

Humoral mechanisms operate on the whole-body level. The system is based on *antibodies,* substances that white blood cells manufacture when they're exposed to a foreign element such as a virus, bacteria, mismatched blood, or a transplanted organ. (Organ transplants are possible because of drugs that inhibit the immune system's response to foreign bodies.)

Antibodies are custom-built to respond to specific threats. They fit like a key in a lock into the chemical structure of the invading organism, neutralizing it. Once produced, antibodies circulate in the bloodstream for up to several years. In addition, once the body has learned the pattern for a particular antibody, it can produce more of them rapidly. Thus, when another attack by the same type of organism occurs, the body can quickly mount a defense. From our perspective, we're usually not even aware of the subsequent invasions; we say that we have acquired "immunity" to the disease.

Cell-Mediated Mechanisms

The immune system also uses *lymphocytes* (white blood cells) to fight off invaders. These cells come in two types: killers and helpers. The killers attack the invading cells and destroy them; the helpers send out chemical signals that bring reinforcements and stimulate the formation of antibodies.

This system of protection is like a finely tuned instrument, and a lot of things can make it go off key. Physical and emotional stress, for example, stimulates production of corticosteroids (natural versions of the steroids used to blunt the immune system in transplant patients). Other factors impair immunity as well, including sleep deprivation, exposure to toxic chemicals, and certain kinds of drugs.

As a master hormone, melatonin helps keep the system in tune at both the humoral and the cell-mediated level. At the same time, it still continues to fulfill its age-old role as a free-radical scavenger, imparting another level of protection *within* individual cells of the body. Although this action isn't usually considered part of the immune system, it serves much the same purpose. By preventing damage to cells, it heads off the deterioration and mutations that can make a cell become cancerous or otherwise susceptible to disease.

Melatonin works on the immune system in a curious and roundabout way. In the absence of any antigens from foreign invaders—i.e., when the body's defenses aren't under attack—the hormone has no apparent effect on the immune system. Rather, melatonin kicks in when the immune system is stressed (whether by infection, cancer, or daily living) and helps restore equilibrium and keep it functioning at optimal levels. In short, melatonin affects the immune system only when the system is "on alert," possibly helping "reset" the immune system after it comes under attack. In this way, melatonin keeps the immune system in good working order and prevents the early appearance of immune-based degenerative diseases. Repeated attacks on the immune system—infections, stress, and so on—may gradually disrupt its rhythms and diminish its effectiveness.

There's ample evidence, from the field and the lab, of the close relationship between melatonin and immunity.

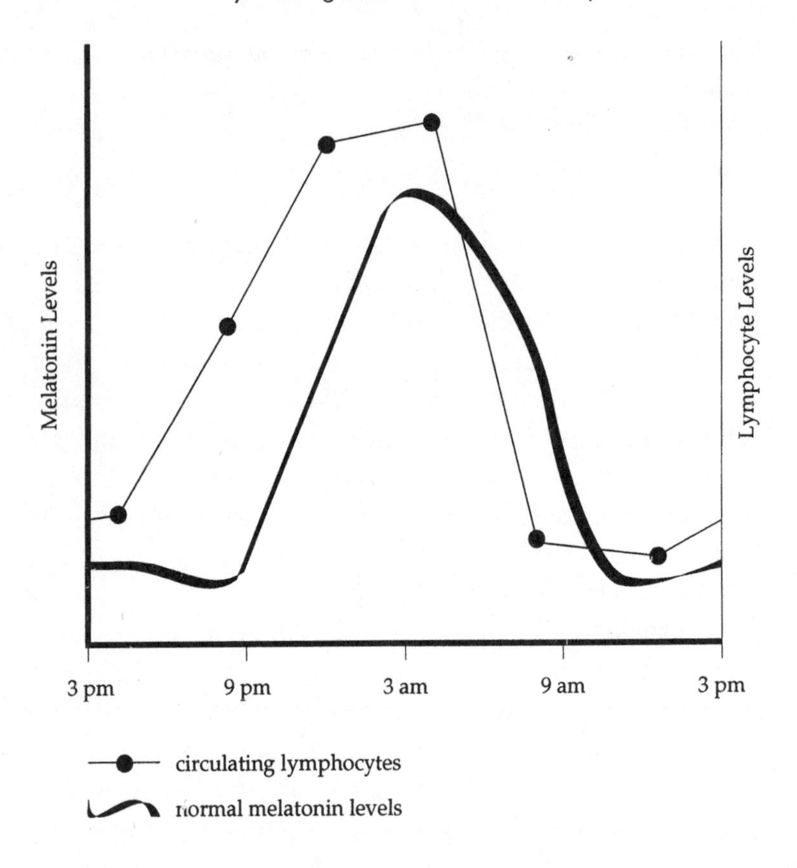

Daily Melatonin Levels Compared with Lymphocyte Levels. *When melatonin levels rise at night, lymphocytes (infection-fighting cells) in the bloodstream rise as well.*

For example, most immune functions follow twenty-four-hour rhythms. They can be thrown off by jet lag and shift work, leaving us more prone to infection. Also, according to one study, we tend to produce more infection-fighting white blood cells when day lengths are longer (suggesting one reason why we might be more prone to flu and colds in winter months). Like other rhythms in the body, these immune rhythms can be "entrained" or reset by melatonin. People who are depressed have been found to have lowered immunity—and, as we've seen, depression is linked to disturbances of melatonin rhythms.

In further studies, melatonin has been shown to have dramatic effects on the thymus, one of the fundamental organs of the immune system. It is here that T-cells—one of the key elements in the body's defenses against invasion—are manufactured. It appears that the thymus undergoes a curious transformation as we age: It grows steadily larger as we approach puberty; then begins to shrivel in upon itself until, in old age, it has virtually disappeared. As the thymus gradually declines, so does our infection-fighting ability. Melatonin seems to protect the thymus and improve its functioning as we grow older.

Additional experiments have shown that melatonin can counteract the effects of stress on the immune system.

Melatonin, Immunity, and Old Age

It so happens that many of the diseases of old age are caused by an immune system that isn't working well. The gradual decline of the immune system explains why we become more susceptible to cancer and infection in old age. But it also explains some other diseases of aging. For example, we now know that at least some types of diabetes and arthritis are caused by irregularities in the immune system: The body fails to recognize its own cells and attacks them as it would a foreign invader.

Thus, melatonin's influence over the immune system is a function of its broader role in controlling the aging process. In fact, medical researchers are beginning to view aging and immunity as two sides of the same coin—with significant implications for future research directions in both fields. For example, it may one day be possible to treat many diseases of aging by manipulating the immune system—and as we'll see in the next chapter, such an approach has already yielded dramatic advances for the treatment of cancer.

In all of this, the role of the pineal gland—and by extension, the role of light—has taken on new importance in our understanding of how we prevent and fight off disease. The day may not be far off when we get a prescription from the doctor that says, "Take ten hours of sunlight, and two hours of darkness, before bedtime."

CHAPTER 6

Melatonin and Cancer

The history of medicine often reads like a mystery novel. A scientist, for example, may observe a curious pattern in which two seemingly unrelated diseases tend to go hand in hand. People in a certain community, or of a certain ethnic background, may turn up dead unexpectedly. Clues are discovered and lost, or filed away as medical curiosities for generations until an additional piece of the puzzle reveals itself. Unexplainable medical mysteries later turn out to be not so mysterious after all. In the nineteenth century, doctors couldn't figure out why so many women who had babies in hospitals later succumbed to "childbed fever." Once they discovered germs, they realized that they themselves were at fault, failing to wash their hands between examinations and spreading infection from patient to patient.

A more modern medical mystery is this: Why does blindness seem to protect women from breast cancer? Although some blind women *do* get breast cancer, a review of 100,000 hospital records by researchers at the Centers for Disease Control indicated that the disease is twice as common in sighted women as in blind women. Pineal researchers believe they have the key to the mystery: melatonin levels. They note that because of the loss of light-dark information to the pineal gland, blindness also affects melatonin cycles, and tends to keep melatonin levels higher than normal. They argue that higher melatonin levels offer some protection against breast cancer. These and other recent studies offer growing evidence of melatonin's ability to protect us from cancer.

Studies dating back to 1940 have shown that a substance produced by the pineal gland could inhibit tumor growth (although nobody knew what the substance was at the time), and that removal or destruction of the pineal gland led to faster growth and proliferation of certain tumors. More recent studies have found that melatonin slows the growth of breast cancer cells in vitro (i.e., in laboratory cultures) and in mice. In hamsters, melatonin slowed the growth of melanoma (a skin cancer) by a factor of five and delayed metastasis (the spread of the cancer to other organs).

Other studies, however, have shown more contradictory results, and some have even shown *increased* levels of melatonin associated with some cancers. These conflicting results may occur as the body produces more melatonin in the early stages of cancer as it attempts to fight the tumor, then decreases its production as the body's defenses are overwhelmed. This may explain the varying clinical and laboratory results, several of which show high melatonin levels early in the disease and low levels among patients in more advanced stages of cancer. Studies of men with prostate cancer and women with breast cancer support this view; in each case researchers found unusually high levels of melatonin in early stages, and a decline in melatonin levels as the disease progressed. Studies are further hampered in that melatonin is measured usually once a day, during the day, ignoring the fluctuation of daily hormone levels.

Despite the confusing and sometimes conflicting results, one clear fact has emerged: When it comes to cancer, melatonin is at the center of the action, and most evidence suggests that it plays a key role in fighting and preventing cancer.

How Melatonin Fights Cancer

Researchers believe that melatonin protects against cancer in at least three ways:

- As an antioxidant, melatonin works directly to neutralize free radicals, which can cause cells to become cancerous by damaging their DNA.
- By stimulating the immune system, melatonin may also help the body kill off cancer cells at their very earliest stages, before they can reproduce and spread.
- As a master hormone, melatonin regulates the production of estrogen, testosterone, and possibly other hormones that slow or prevent the growth of some kinds of tumors.

In other words, melatonin appears to work on one level to prevent the formation and spread of cancer cells, and on a more specific level against certain kinds of tumors—especially tumors related to the reproductive system. Most research into melatonin as an anticancer agent now focuses on these kinds of tumors.

Melatonin and Breast Cancer

Some cancers of the breast and reproductive organs are particularly sensitive to levels of estrogen, the female hormone. These cells contain "receptors"—chemical structures that can recognize estrogen. In such estrogen-receptor-positive (ER-positive) tumors, estrogen released by the body's reproductive organs makes them grow faster. Approximately 70 percent of breast cancers are ER-positive.

As medical researchers begin to understand the role of these sex hormones in the promotion of cancer growth, they have developed new types of treatments. One of the most significant breakthroughs in the treatment of breast cancer, for example, is tamoxifen and related drugs. These drugs act as "antiestrogens," blocking the effects of estrogen and so helping slow the rate at which ER-positive tumors grow. Studies are now under way to determine whether such drugs can also *prevent* such cancers from forming in the first place.

The discovery of estrogen's role in promoting these tumors suggests that melatonin may be a useful part of the treatment regimen. As part of its role as the body's timekeeper, melatonin affects the reproductive cycle. In seasonally breeding animals, in fact, changing melatonin levels control reproductive rhythms and determine when the animals go into heat. It has similar, though more subtle, effects on human reproductive cycles as well; in fact, humans are more fertile in certain seasons.

Melatonin's influence over the reproductive hormones has sparked considerable interest in its potential therapeutic uses. For example, it's now being evaluated as an aid to contraception (see chapter 16). Similarly, its ability to influence reproductive systems may make it an effective treatment for hormone-dependent cancers, and several clinical studies are under way to explore its ability to enhance other anticancer drugs.

In men, certain kinds of prostate cancer are similarly stimulated by testosterone. Here, too, melatonin has the potential to protect against these cancers by regulating testosterone production.

The melatonin connection may help explain a longstanding mystery about breast cancer: its seasonality. Melatonin levels are generally higher in the wintertime (at least in the northern latitudes) as a consequence of day length. More cases of breast cancer are detected in late spring and early winter, when melatonin levels are lower. In fact, melatonin levels may one day serve as an early predictor of some kinds of cancers, since women with ER-positive breast cancer and men with testosterone-dependent prostate cancer have lower levels of melatonin than those without the disease or with benign tumors.

Several experiments are now under way to determine if melatonin can be used to prevent or slow the progression of breast cancer. At Tulane University School of Medicine, scientists grew human breast cancer cells in the laboratory and then added melatonin to some of the cultures. The results: ER-positive cells grew only one-fourth to one-half as quickly as untreated cells. Cancer cells without estrogen receptors weren't affected by melatonin. Other studies have shown that when melatonin is combined with tamoxifen, it enhances its effectiveness against cancer cells cultured in the laboratory.

It remains to be seen how such a combination might work in human patients. In the Netherlands, a large study of melatonin's potential to prevent breast cancer was launched in 1991, but it will take at least a decade to complete. But these early findings suggest that one day melatonin may be part of the standard arsenal used to treat and prevent some forms of breast cancer, and that it may enhance the properties of other anticancer agents, making it possible to use less toxic doses.

Other Cancers

Cancer of the endometrium—the lining of the uterus—is another hormone-dependent cancer, and it too may be linked to abnormalities of melatonin production. The evidence is circumstantial but significant: Endometrial hyperplasia, a change in this tissue that many believe is a precursor to cancer, is diagnosed more frequently in the wintertime. Also, anovulation, the lack of ovulation, which is also a risk factor for endometrial cancer, is more common in winter months. Further, the risk of endometrial cancer rises during and after menopause, when melatonin production drops off dramatically. And finally, obesity and diabetes, additional risk factors for endometrial cancer, are also associated with impaired melatonin rhythms.

Melatonin deficiencies have been linked to other cancers as well, including Ehrlich's tumors, sarcomas and fibrosarcomas, and melanoma. Melatonin has reportedly improved symptoms and extended survival times for patients with lung cancer, stomach cancer, bone cancer, and cervical cancer.

Melatonin, IL-2, and Cancer

In addition to its effects on ER-positive cancers, melatonin is also being studied as an adjunct to a new and radically different approach to treating cancer.

Until very recently, there were only two basic approaches to cancer treatment: Doctors could attempt to remove the cancerous cells from the body, or kill them within the body. Neither approach is particularly subtle, and both of them have drawbacks. When cancers are surgically removed, it's hard to be sure all of the cancerous tissue is taken out, and it's often necessary to take out a lot of normal tissue with it. Some cancer cells can be killed with chemotherapy or radiation, but these treatments take a toll as well. They kill not only cancer cells but normal cells.

Beginning in the 1980s, cancer researchers began exploring new ways to help the body fight cancer. One approach, as we've seen, is with hormonal treatments. Another is to stimulate the immune system. Most cancer researchers now believe that cancer represents a failure of the immune system. Cancers are cells that have mutated; that is, their DNA has changed in a way that makes them grow uncontrollably. But with billions upon billions of cells in our bodies, such mutations must occur with alarming frequency. They may be caused, for example, by exposure to the earth's natural radiation and to free radicals, as well as by simple errors that creep into the cell's genetic code when it reproduces. Many of these mutated cells, if left unchecked, would eventually grow into tumors.

So what stops them? Is life an extended game of Russian roulette, with our chances of getting cancer governed by how long we've been playing the game? On the contrary, there's evidence that a vigorous immune system actually recognizes and destroys these cancerous cells, much as it does harmful viruses and bacteria. It seems that the issue isn't so much whether your body houses cancerous cells as what it does about them.

This idea launched a new approach to treating cancer: immunotherapy. By stimulating the immune system, scientists hoped they might be able to treat cancer. Evidence of a link between the immune system and cancer was confirmed by clinical trials of interleukin-2 (IL-2), a substance produced by the immune system. IL-2 seems to function something like the old airplane spotters who watched the skies during World War II: It helps the immune system "spot" cancer cells by homing in on them.

Early experiments with IL-2 were dramatic. However, they had serious drawbacks: IL-2 tends to make many patients violently ill. In fact, in the initial studies an alarming number of subjects died of the side effects of treatment.

These early studies involved patients with otherwise untreatable cancers, so even with the drawbacks IL-2's effectiveness represented an important advance over the then-current state of the art. But at the same time, the successes have been inconsistent: Sometimes IL-2 is effective and sometimes it's not. IL-2 has proven most effective in reducing tumors in metastatic kidney cancer, colorectal cancer, melanoma, and lymphomas.

Experiments with IL-2 have generally found its usefulness limited to renal (kidney) cancer and melanoma (a malignant skin cancer). A group of researchers at San Gerardo Hospital in Monza, Italy, conducted tests to see if melatonin could enhance the effectiveness of IL-2 against other cancers as well. They administered low doses of IL-2 (to reduce the severe side effects) with melatonin to eighty-two patients, most of whom had metastases—that is, cancer that had spread from its initial site to distant organs. The regimen shrank the tumors in twenty-one of the patients, and produced complete remissions in four patients. Side effects were mild in all the patients. Another experiment by the same group found that melatonin significantly improved the effectiveness of IL-2: 7 percent of patients receiving both melatonin and IL-2 experienced complete remissions, versus none in the group receiving IL-2 alone. Twenty percent of those receiving the combination therapy had partial remissions, versus 3 percent of those who received only IL-2. Survival after one year was 46 percent for the first group, versus only 15 percent for the IL-2-only group.

IL-2 and melatonin seem to work hand in hand. Not only does melatonin improve IL-2's effectiveness, but IL-2 also enhances melatonin production in cancer patients. For example, a study of seven patients with advanced lung cancer found that they completely lacked the normal rhythm of melatonin production. After receiving IL-2, four of them reestablished a nighttime melatonin peak.

A Caveat

Though early research into melatonin's cancer-fighting qualities is promising, there are many still-unanswered questions. *If you or someone you know has cancer, don't add melatonin to your treatment regimen without consulting your doctor.* Beyond the few clinical trials now under way, there are more questions than answers about how melatonin affects cancer. One researcher, for example, points out that with other bodily processes such as reproduction, melatonin sometimes stimulates them and sometimes inhibits them, depending, for example, upon what time of day it's administered. Similarly, studies in mice found that melatonin administered in the late afternoon slowed the growth of tumors, but melatonin administered in the early morning seemed to actually *stimulate* the growth of existing tumors. In particular, people with Hodgkin's disease, leukemia, lymphoma, or multiple myeloma shouldn't take melatonin, but for anyone who has cancer, it's vital to follow your doctor's prescribed treatments and not initiate any changes in your treatment on your own. In most cases, we simply don't yet know enough to say what melatonin's role should be in the treatment of individual cases of cancer.

CHAPTER 7

Melatonin and EMFs

There's been a lot of publicity in recent years about electromagnetic fields (EMFs). These invisible fields of charged particles are a by-product of our modern electric lifestyle. Dubbed "currents of death" in one best-selling book, they may be responsible for untold cases of cancer, depression, and other diseases, according to some critics.

Others, however, argue that such alleged effects are pseudoscientific hogwash, the studies are poorly designed and the results inconsistent. Unlike other alleged carcinogens, EMFs don't seem to be associated with only one or a few kinds of cancer; rather, they seem to be vaguely related to a diverse range of cancers. Besides, EMFs are far less powerful than the body's own electrical signals; how, then, could they affect body cells and trigger cancerous changes?

Our new understanding of the pineal gland may be the missing part of this puzzle, because it has become clear that EMFs do have dramatic effects on the pineal gland and melatonin production. And as we learn more about melatonin's role in the body, we are finding that many of the ill effects claimed for EMFs are the *same ones that occur from low melatonin levels*. Like sunlight, EMFs don't act directly on most human cells, but indirectly by way of the eyes and the pineal gland. The evidence is growing that the health effects of EMFs are real, and are intimately related to the fact that EMFs impair the pineal's production of melatonin.

What Are EMFs?

The term "electromagnetic field" is probably too broadly defined. It's used loosely to describe several different forms of energy. Though these phenomena are related, they're not identical—nor are the health effects of each.

The first of these, *electromagnetic radiation,* isn't really a "field" at all. It's a burst of electromagnetic waves. Strictly speaking, electromagnetic radiation encompasses everything along the *electromagnetic spectrum,* ranging from extremely low frequency waves used in military communications to microwaves and X rays. Human beings have built-in sensors that can detect electromagnetic radiation in certain wavelengths. We call it visible light.

There's no difference between light, microwaves, and X rays except for how long or short the wavelengths are. The shorter the wavelength, the higher the energy the waves contain.

In recent decades there's been a lot of interest in the

health effects of microwaves because of the many sources of exposure to them. Microwave towers and satellite uplinks at television stations, for example, are potent sources of microwave radiation. Radio and television transmitters also emit electromagnetic radiation, though in a slightly different part of the spectrum.

In very close proximity, these waves of energy can be extremely harmful, even deadly, because they heat up objects in their path (that's the principle behind microwave ovens). There are also concerns about more subtle long-term effects on people who live near radio and television transmitters, microwave towers, and other sources of electromagnetic radiation.

Electric fields are fields of electrically charged particles that are given off as a by-product of electric current. *Magnetic fields* are similar fields of energy surrounding magnets. The key difference between electric and magnetic fields involves their charged particles. In electric fields, these particles are in motion, giving them the ability to create electric currents in materials they come into contact with. In magnetic fields, the particles are stationary. You may remember elementary school science experiments in which iron filings sprinkled around a magnet arranged themselves into arcing patterns from one end of the magnet to the other. Those arcs were visible evidence of magnetic fields. The earth is itself a big magnet, and we all live within the magnetic field it generates.

Electric power lines, transformers, motors, and the like give off both electric and magnetic fields. There are some significant differences between the two. For one thing, it's fairly easy to shield electric fields; many ordinary materials reduce their impact. Magnetic fields, by contrast, tend to pass through steel, brick, masonry, wood—in fact, virtually all ordinary materials.

Some evidence seems to suggest that electric *currents* may be the ultimate culprit in EMF exposure. When electrically conducting material—be it copper wire or human tissue—is placed within an electric field, the field causes current to flow in the material.

Magnetic fields can also induce current, but only if they're *moving* in relation to the material. For example, if you turn your head from side to side, it changes position relative to the earth's magnetic field. The currents induced in this way are both random and minuscule, and seem to have no effect on human biology (at least as far as we know). However, animal experiments show that *rapidly changing* magnetic fields affect the pineal gland, whereas more slowly changing magnetic fields have no effect. Again, the key difference seems to be that rapidly changing magnetic fields cause electric current to flow.

The currents generated by typical EMF exposures are so low that our senses can't detect them. Laboratory studies suggest that EMFs don't disrupt cells the way a chemical agent or strong radiation does. One of the objections to theories of EMF damage is that such currents are far less powerful than the constant flow of electricity that the body produces as it goes about its daily business. Such small changes would be lost in the "background noise" of the body's own electrical activity.

In addition, researchers have been troubled by the variety and inconsistency of reported damage: In some cases, EMFs are alleged to cause childhood leukemia. Other studies find potential problems with breast cancer.

The Pineal Connection

As we learn more about the effects of EMFs on the pineal gland, some of these mysteries are beginning to clear up. There's a compelling body of research showing that EMFs interfere with melatonin production. And as we saw in chapter 5, low melatonin levels can cause immune deficiencies. Even subtle impairments of the immune system can generally increase the risk of cancer, and (just as the EMF research shows) you wouldn't necessarily see any clear pattern of a single type of cancer. At the same time, reduced melatonin would inhibit the ability of cells to repair damage caused by other factors, again increasing the risk of cancer. Here, too, the development of cancer would be inconsistent, since it would also require another risk factor besides EMFs.

In other words, it may be that EMFs don't *cause* cancer directly, but, by reducing melatonin production, leave us more *susceptible* to cancers caused by other factors. That would explain why the relationship between EMFs and cancer has been so hard to pin down. It also would explain how EMFs could promote cancer without causing direct damage to the body's cells.

The effects of EMFs on the pineal gland have been documented in animals and humans. In a series of studies that stretched over several years, for example, rats were exposed for twenty hours a day to 60-hertz electric fields—the same frequency produced by ordinary household appliances. After thirty days' exposure to the fields, the rats' melatonin production was virtually shut down. After the animals were no longer exposed to the fields, melatonin production shot back up to normal—typically, within three days.

Another experiment in 1986 showed similar results: After three weeks of EMF exposure, nighttime melatonin production in rats was only about half of normal levels; after four weeks it was only a third of normal levels. When rats were removed from the electric fields, their melatonin levels also returned to normal.

Epidemiological studies, which examine patterns of disease incidence, also support the link between EMFs and diminished melatonin production by demonstrating a relationship between EMF exposure and the kinds of health problems that occur from low melatonin levels.

EMFs and Hormone-Dependent Cancers

As we saw in chapter 6, melatonin is believed to play a role in preventing hormone-dependent cancers from forming. And since EMFs can diminish melatonin production, you'd expect to see more such cancers among people who are exposed to EMFs. Researchers are just beginning to explore the relationship between EMFs and hormone-dependent cancers, and it's too early to draw any definite conclusions. But there are some intriguing indications of a potential link. Male breast cancer was found to be disproportionately high among a group of New York telephone workers who were probably exposed to high and chronic levels of EMFs as a result of working near power lines.

Another study found up to a sixfold increase in cases of male breast cancer among workers whose jobs exposed them to high levels of EMFs.

EMFs and Depression

There's a strong relationship between melatonin and mood (see chapter 16). In light of that fact, you'd expect to find that people exposed to EMFs are more prone to mood disorders such as depression. At least two studies support such a link. In 1988, a study of people living in housing projects in Britain found that people living in apartments where EMF levels were high were also more likely to be depressed. A 1979 study found higher rates of suicide among people living near overhead power transmission lines.

There are a number of ways in which EMFs could potentially affect moods. By reducing melatonin levels directly, they could alter the brain chemistry in ways that make it more prone to depression. In fact, several studies have found lower melatonin levels in depressed patients. Or by altering pineal production they could induce emotional disturbances by interfering with normal sleep patterns. Such disruptions (such as insomnia and daytime napping) are often associated with depression.

The Eyes Have It

Most researchers feel that EMFs probably affect the pineal by way of the eyes. The retinas, of course, are extremely sensitive to light, which is a form of electromagnetic radiation. Thus, it's not unreasonable that they would also be sensitive to electromagnetic *fields*—EMFs—even if we can't consciously perceive these forces at ordinary levels of exposure. (In fact, strong magnetic fields can produce flickering or shimmering visual effects.)

It's not clear why our eyes would be sensitive to these fields. For most of the millions of years of human evolution, we were exposed to fields of such strength only during electrical storms. And, as we've seen, the fields generated by the earth's magnetic field are much weaker. This sensitivity could be an evolutionary leftover—a remnant of some ancestral system that's no longer functional in humans. Birds probably use magnetic fields to help navigate during their annual migrations, and the pineal gland may be attuned to respond to these changes in ways we can't yet detect.

What's Safe?

Every day, we're exposed to countless sources of EMFs: household appliances, office machinery, fluorescent lights, power lines, household wiring, and others. How dangerous is this exposure?

The answer is that we just don't know. Rates of certain kinds of diseases have risen dramatically in the last hundred years—a period of time corresponding to the rise of electrical power and exposure to EMFs. But the world in which we live has changed in many other ways as well, and it's hard to sort out which of these health problems might be caused by EMFs as opposed to, say, pollution, lifestyle changes, and so on.

Despite such difficulties, experts have suggested an average exposure of 2 milligauss as a threshold for safety from EMFs. The threshold is arbitrary; it's based more on "typical" levels of exposure than any clear-cut evidence of health effects. In other words, most people are exposed throughout the day to EMFs in the 2 milligauss range; anything above 2 milligauss indicates greater-than-normal exposure. (The table on pages 151–152 shows exposures from a variety of common EMF sources.) There's nothing to say, however, that these "typical" levels are harmless, or that higher levels are more dangerous.

"Prudent Exposure"

Faced with these difficulties, what's the best approach to managing your EMF exposure? Researchers at Carnegie Mellon University recommend an approach called "prudent exposure," and they suggest that it offers the best way to address the issue of EMFs—at least until we know more about their specific health effects.

This approach relies on common sense and suggests that we look first at making simple changes that will reduce our total EMF exposure. For example, replacing dimmer switches in your home or repositioning your computer monitor may do more to reduce your EMF exposure than selling your house because you live near power lines.

Think of EMF control as you would sun exposure. Just because UV rays are harmful doesn't mean you can't go to the beach; rather, it means you should use *reasonable* precautions such as sunscreen and a hat. Take the same approach with EMFs. In chapter 12, we offer such an approach to identifying and reducing such sources of exposure.

In addition, we can use our knowledge of the link between EMFs and melatonin to protect ourselves. For example, as we'll see later, if you must live or work in an environment that has a lot of EMFs, consider taking melatonin supplements. In addition, you can limit the effects of EMFs by adopting other strategies to enhance your body's natural production of melatonin, such as increasing your exposure to sunlight and maintaining regular schedules of activity.

As you can see, melatonin plays a surprisingly vital role in many aspects of our health. Perhaps equally surprising, by making some simple and healthy changes in the way you live, you can begin reaping its benefits today.

In the pages that follow, you will find ways to stimulate your body's own production of melatonin. Chances are, you'll begin to notice the differences immediately: You'll feel more rested, more energetic, more alert, more *in tune* with your body's natural rhythms. The program is an effective antidote to many of the man-made stresses of modern life, and will bring you closer to a way of life that the human body is designed for.

And yet most of these changes are marvelously simple. They're not a hardship. They don't require a lot of willpower. In fact, as you begin to feel more *rested* and energetic, you'll *want* to follow the program. And as these changes improve your daily life, remember: They're also helping you *stay* healthy as you grow older.

Sound good? Then let's get started.

Part Two

THE MELATONIN METHOD

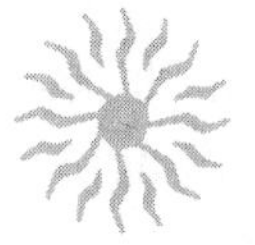

CHAPTER 8

The Melatonin Lifestyle Inventory

In the preceding chapters, we've looked at melatonin's profound effects on health and longevity. But the really good news is that unlike many medical breakthroughs, you don't have to wait years and years to see practical benefits. You don't need to get a prescription from your doctor. You don't have to wait for your insurance company to cover it. You can start putting melatonin to work for you *today* by making changes in your lifestyle that will boost your body's natural production of melatonin and, if you choose, by adding to this supply with oral supplements for only a few dollars a month.

The Melatonin Method is a simple, practical approach to making these new findings work for you. Here we'll show you how to make changes that will bring your lifestyle into closer harmony with your body's rhythms and processes.

As you work through the chapters that follow, keep in mind that *quality* is more important than *quantity*. If your body is producing lots of melatonin, but at the wrong times, you're missing out on much of its benefits. In fact, it may do more harm than good. Symptoms of depression and jet lag, for example, occur when your body's getting too much melatonin, or producing it at the wrong times.

The Melatonin Lifestyle Inventory that follows will give you a general sense of where you stand in terms of melatonin production. Then, in the following chapters, we'll explore lifestyle in more detail and offer some changes that can help keep your melatonin rhythms on track. You'll find that most of these changes are simple and easy. One of the good things about a melatonin-friendly lifestyle is that it doesn't require a lot of sacrifices or disruption. In fact, most of the changes will help you feel better immediately.

Mostly you need an understanding of how our bodies make and use melatonin. Once you look at your lifestyle from this new perspective, you'll quickly find simple and commonsense strategies that will help you work *with* your body's rhythms, not against them.

In the laboratory, researchers use blood tests to measure the amount of melatonin that's produced and circulating in the body. But unless you're willing to submit to a needle every two hours each night, it's impractical to measure your melatonin levels directly. However, you can get a good idea of whether you're getting enough melatonin by looking at how you live and how you feel. In the chapters that follow, we offer a series of questionnaires that explore these issues in more detail to help you identify whether and how you're at risk and suggest practical strategies for maximizing melatonin's life-giving benefits.

Melatonin Lifestyle Inventory

For each of the following set of statements, select the one that most closely corresponds to your lifestyle. Then refer to the discussion that follows to learn about how your lifestyle may be affecting your melatonin level.

1.

_____ In general, my life follows predictable patterns of activity; I tend to do the same kinds of things at about the same time every day.

_____ My life is generally pretty hectic.

2.

_____ I spend a lot of time outdoors.

_____ I spend most of my work and leisure hours indoors.

3.

_____ I rarely use medications.

_____ I use medications (prescription and over-the-counter) regularly.

4.

_____ I feel that my life is in tune with my body's rhythms.

_____ I feel that I don't "fit" into my current lifestyle very well.

5.

_____ I usually keep stress under control.

_____ I lead a stressful lifestyle.

6\.

_____ I often have trouble sleeping.
_____ I'm a good sleeper.

7\.

_____ I'm usually alert during the day.
_____ I often feel drowsy during the day.

8\. I am:

_____ 18–35 years old
_____ 36–50 years old
_____ 51–70 years old
_____ 70+ years old

9\.

_____ I hardly ever get depressed.
_____ I often feel depressed.

10\.

_____ When everyone else is sick, I usually remain well.
_____ I seem to catch every cold and flu that's going around.

11\. I have a family history of, or suffer from, the following health problems:

_____ Diabetes
_____ Heart disease
_____ Cancer
_____ Multiple sclerosis

12\. I live or work in an environment that may expose me to high levels of:

_____ Electromagnetic fields (EMFs)

_____ Air pollution
_____ Industrial chemicals
_____ Radiation (including X rays)

13. My diet
_____ includes lots of fresh fruits and vegetables
_____ includes a lot of fast foods and processed foods

14. I exercise
_____ Moderately
_____ Strenuously
_____ Not at all

Interpretation

1. A lifestyle that follows predictable routines will make it easier for your body to maintain its rhythm of melatonin production. Periods of light and darkness set the primary rhythm, but other environmental cues exert powerful influences as well. See chapter 9 for strategies on how to keep your schedule synchronized with your body's basic rhythms.

2. Humans evolved in a world governed by the sun, but modern conditions often divorce us from this most potent timekeeper. Many health problems, including depression, sleeplessness, and such physical manifestations of stress as heart disease and high blood pressure may be brought about by our lack of exposure to sunlight. See chapter 9.

3. Some drugs and medications interfere with our

bodies' ability to make and use melatonin. Even common and seemingly innocuous over-the-counter remedies such as ibuprofen may depress the functioning of the pineal gland. See chapter 13.

4. If you feel that your body is rebelling against the way you live, listen to it. The strategies outlined in the following chapters can make big changes in the way you feel. In fact, by making minor lifestyle changes now, you can prevent the kinds of problems—high blood pressure, for example—that will require much more radical lifestyle changes later on.

5. Some people think of stress as a sign of success, and they wear it almost as a status symbol. In fact, chronic stress is almost *always* bad for you. In our early ancestors, it was a short-term response to physical threats; ongoing stress, however, diminishes our effectiveness over time by short-circuiting melatonin's anti-aging and immune effects. It makes us old and sick.

6. If you have trouble sleeping, it's a strong signal that your melatonin rhythms are out of sync. By incorporating the melatonin-friendly lifestyle changes in chapters 9 and 10, you take advantage of your body's natural sleep-wake patterns to get more restful sleep.

7. Sleepiness during the day, like sleeplessness at night, suggests that your basic circadian rhythms may be out of sync. In addition, it may be a sign of depression.

8. If you're young, congratulations—your melatonin levels are probably high. The Melatonin Method will help you keep them that way. If you're older, you may already have noticed some of the signs of reduced mel-

atonin levels: trouble sleeping, reduced resistance to disease, etc. Lifestyle changes and, perhaps, melatonin supplements can help delay these and other signs of aging.

9. Depression—especially when it comes and goes with the seasons—has been linked to abnormal melatonin rhythms and levels. If you suffer from depression, see your doctor. If you're being treated for depression, ask your doctor about phototherapy as part of your treatment.

10. Melatonin plays a key role in keeping the immune system strong. If you catch a lot of colds, low melatonin may be to blame. See chapter 5.

11. It's not known whether low melatonin levels run in families, but many diseases that are known to be related to melatonin production do. If you have a family history of diabetes, heart disease, cancer, or multiple sclerosis, improving your melatonin levels may reduce your risk.

12. Environmental factors such as pollution and EMFs can damage your melatonin levels. Chapters 11 and 12 suggest ways to protect yourself.

13. Diet can help or hinder your melatonin production. Some foods, such as those containing tryptophan, may give your body the basic building blocks it can use to manufacture melatonin. Foods high in antioxidants help enhance melatonin's antiaging properties. See chapter 10.

14. Moderate exercise can improve your immune system, help you sleep, and keep you fit. However, ex-

treme exercise may actually reduce your melatonin by releasing free radicals into your bloodstream. See chapter 11.

This assessment is designed to give a broad view of your lifestyle and help you gauge how well your habits fit with what we know about melatonin production and use. In the chapters that follow, we'll explore key areas in detail, and suggest areas in which you can begin to make some changes.

Keep in mind that there are no absolutes. Our bodies evolved in an environment very different from the way most of us now live, and in exchange for the conveniences of modern life, we must inevitably make some compromises with what would be ideal from the standpoint of melatonin production. But we should make changes where we can, for the long- and short-term benefits will enhance our lives.

Once you've completed this inventory, take a few moments to consider what it reveals. It may suggest tackling one part of your lifestyle first—for example, your work schedule or your exposure to EMFs. In the end, however, the changes suggested in the chapters that follow are cumulative, and the more of them you can implement, the better.

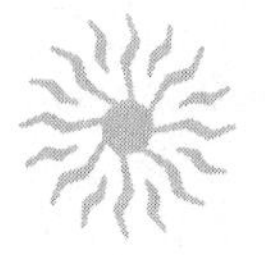

CHAPTER 9

Keeping in Sync with the Sun

Civilization has its good points and its bad points. On the one hand, those of us who live in modern, industrialized countries are relatively free of some of the most terrifying diseases of ages past: smallpox, polio, cholera, various plagues. On the other hand, there's a host of new things to worry about: cancer, heart disease, ulcers, stress itself. To a large extent, these diseases are by-products of the way we live. The more industrialized a society, the higher the rates of these "lifestyle diseases."

The reasons for this shift are complex and not fully understood. Some researchers feel that there are more lifestyle diseases today simply because other diseases aren't killing as many people as they once were. In support of this view, which might be called the "you-have-to-die-of-something" school, is the fact that such diseases tend to strike later in life.

Other studies, however, have implicated various features of modern living, such as diet, industrial toxins, and stress. If, as we now suspect, melatonin controls the aging process, we will have to add another culprit to the list: a lifestyle that becomes less and less in tune with natural rhythms of light.

Alarm clocks, light bulbs, shift work, and late-night television are all symptoms of our nonstop twenty-four-hour society. As attention has focused on the link between the pineal gland, melatonin, and aging, researchers are beginning to look at whether these and other disruptions in circadian rhythms might play a role in the development of lifestyle diseases. It's too early to say for sure, but given melatonin's central role in protecting the body's cells from damage, there's a good chance that disruptions in our natural circadian rhythms do leave us more vulnerable to these diseases. That's not to say, of course, that disruption of the natural light-dark cycle is the *only* thing we need to do to stay healthy. We can't use it to offset the harmful effects of smoking, fatty foods, lack of exercise, and other high-risk factors. But it may enhance the benefits of other healthy lifestyle changes.

In this chapter, we'll offer a practical approach to keeping your internal clock in sync with the sun. We'll help you identify factors that may be throwing this rhythm off, and then suggest simple ways to make your schedule and your routines more melatonin-friendly. Finally, we'll give you some tools to help you determine what differences these changes are making in your life.

These simple changes to your daily routines can offer both short-term and long-term benefits by helping your pineal gland produce the right amounts of melatonin at the right times. As you begin to align your rhythms with those of the sun, you're likely to find yourself more rested, more energetic, and more alert. Because the pineal gland also helps regulate the immune system and moods, you may also discover that you're less susceptible to colds, flu, and other infections, and less prone to depression and mood swings. And over time, these and other melatonin-enhancing strategies may also help mitigate the effects of such health problems as high blood pressure, memory loss, arthritis, and other age-related conditions.

Nature's Clock

Life was a lot simpler in the good old days. Before all the modern conveniences, humans were more or less forced to live by the sun. They got up when it got light out and spent a good part of their days working outdoors. When the sun went down, they slept: With eyes ill-suited for darkness, they couldn't hunt and they couldn't plant. Artificial lights like fires and lanterns burned precious fuel; they were a fire hazard and they could attract wild animals and other unwelcome visitors. Day in and day out, our bodies followed the natural rhythms of the sun. We worked longer hours in the summer, when the light lingers into the evening, and in the wintertime we went to bed early. These daily and seasonal rhythms governed our lives.

These days, the rhythms are turned topsy-turvy. Some nights we go to bed early; others we stay up until the wee hours. Year-round, we work at jobs where the hours are set by the clock, not by the light in the sky. (If anything, we tend to work longer hours in the winter.) We set alarms in the morning to wake ourselves up while it's still dark. We scurry off to work and school, where we spend our days under artificial lights. Twice a year we reset our clocks to daylight or standard time and wait for our bodies to catch up. Some of us have jobs that require us to reset our biological clocks every few weeks as we move to a new shift. Others work the night shift permanently, their periods of sleep and activity totally out of phase with the sun.

Throughout all of this, our pineal gland tries hard to keep to the rhythm it knows best, the rhythm that was established by millions of years of evolution. But when it misses its cues—darkness at bedtime, sunlight when we rise, regular rhythms of sleep and wakefulness—the production of melatonin is diminished.

Step 1: Charting Your Rhythms

To begin to get an idea of how your lifestyle might be affecting your body's production of melatonin, chart your basic rhythms and compare them to natural cycles of light and dark. The charts here are similar to the ones used in studies of circadian rhythms in laboratory animals. These rhythms directly influence melatonin production in the pineal gland. Below, we'll show you how to compare your rhythms with ideal patterns, and how to make changes to bring those rhythms more into line with natural periods of light and dark.

How to Use the Chart

Make several copies of the blank chart that follows. Use the chart to track two weeks of activity at a time. Complete the "baseline" chart now; complete the second chart after you've made some of the lifestyle changes described below. You may also wish to complete a chart in different seasons to identify any seasonal changes, or at other times if your schedules change dramatically. Keeping the chart is simple and takes only a few minutes a day:

1. Every day for the next two weeks, note what time you go to bed and when you get up. Use the *actual* times; if you oversleep one morning, use the time you actually got up, not when your alarm clock went off.

2. Chart, hour by hour, your exposure to daylight during the day. For each day, note the time of your first exposure to daylight (whether it's when you open the blinds, step outside for the paper, or leave the house in the morning). Exposure to just artificial light doesn't count; if you work exclusively under fluorescents all day, record it as dark time. However, any exposure to daylight (whether it's sunny or not) counts. So if you sit near an outside window at work, chart it as daylight exposure.

3. (Optional) For a comparison with natural daylight rhythms, you can chart times of sunrise and sunset. Look in the newspaper to find the times. Mark the periods of day and night at the bottom of the chart. (Use the same times for each day on the chart; the times change only by a few minutes over the course of two weeks.)

4. (Optional) If you want more detail in your chart, you can also chart your activity levels throughout the

day. Once an hour, mark your average activity level for the preceding hour (high, medium, or low).

Baseline Activity Levels

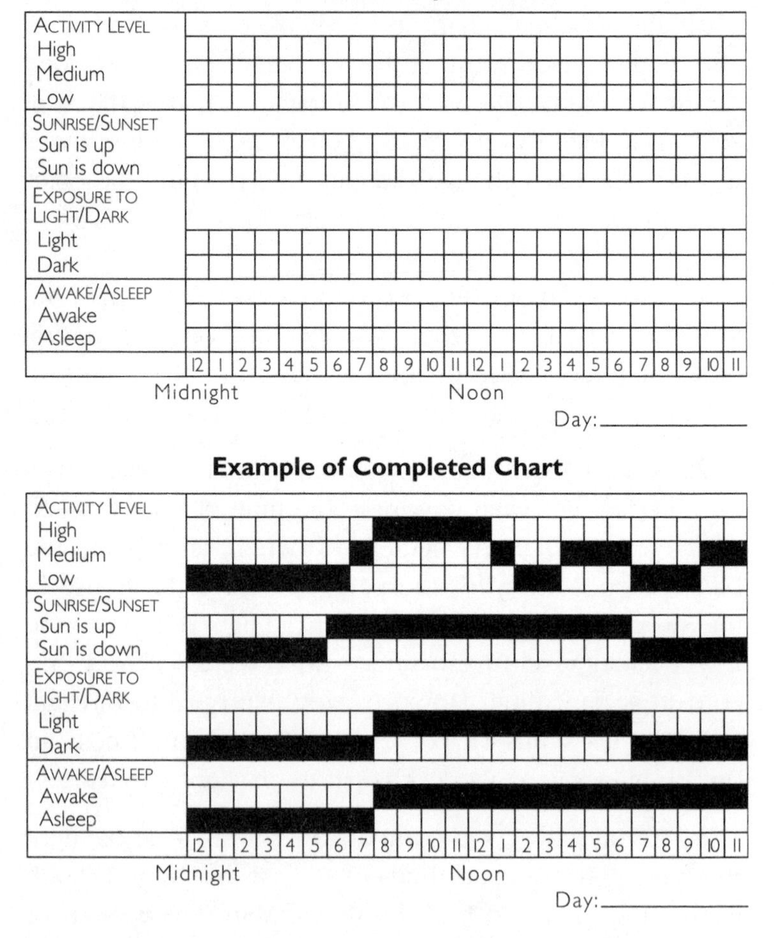

Interpreting Your Circadian Rhythm Chart

The patterns revealed in your circadian rhythm chart can give you insights into your body's rhythms. Find the chart below that looks most like yours:

Ideal Patterns of Activity

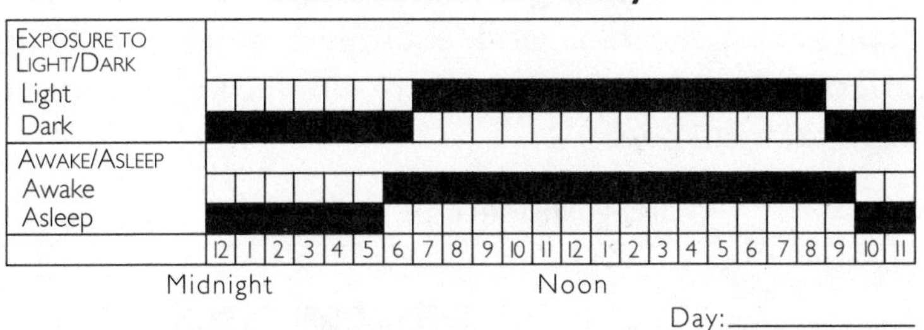

Ideal. This pattern indicates that your body's rhythms are closely synchronized with natural daylight. Sleep comes shortly after nightfall, and activity begins shortly after the sun comes up. This rhythm means that your body has the opportunity to produce large amounts of melatonin, and to do so over a long period of the night. This pattern creates a large difference between daytime melatonin levels (low) and nighttime levels (high), which helps to keep your body well synchronized.

How much you sleep isn't as significant as *when* you sleep. There's a lot of variation in the number of hours of sleep people need in order to feel rested: A few of us function well on only four or five hours a night; others need ten or more. Most of us fall somewhere in between. What *is* important is whether this sleep is synchronized with the patterns of light and darkness.

Phase-shifted Pattern of Activity

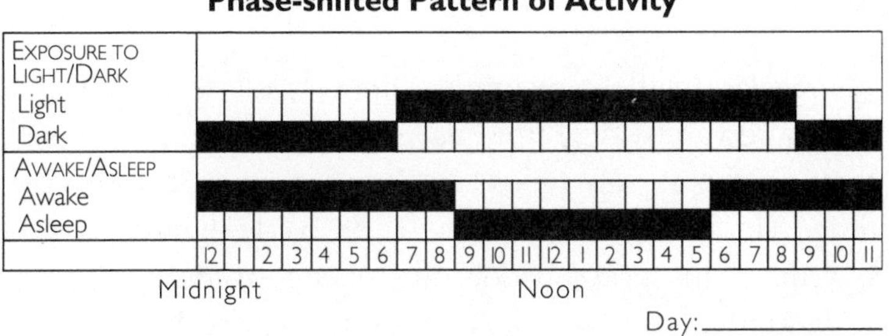

Phase-shifted. If your pattern looks like this, you're suffering from the biological equivalent of jet lag. Your sleeping rhythms are out of step with the visual cues of light and darkness.

Irregular Patterns of Activity

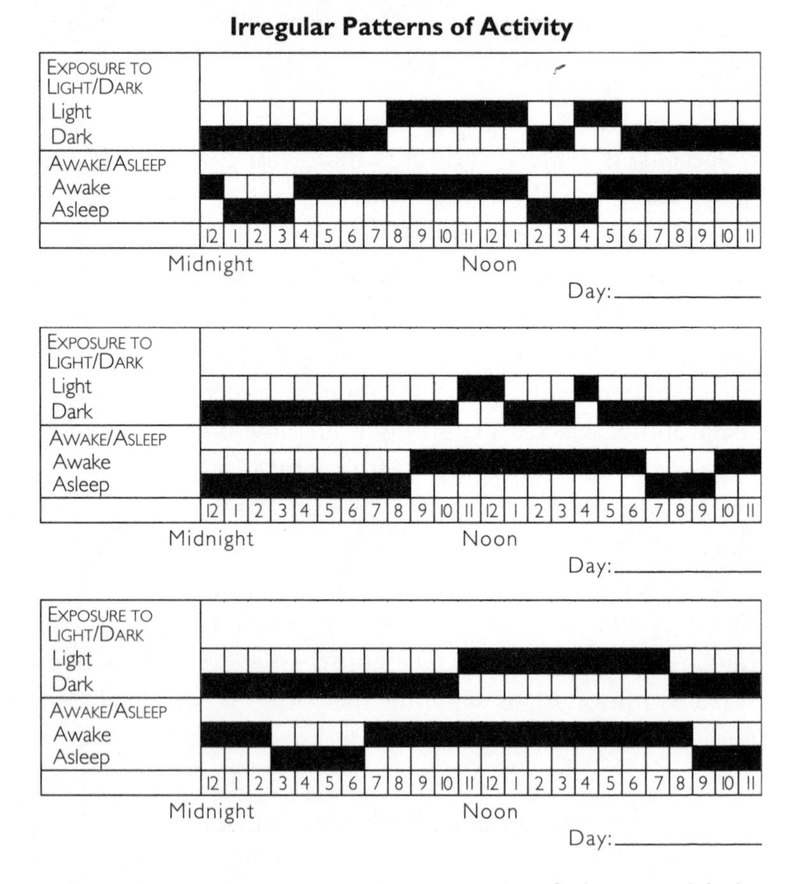

Irregular. In these examples, periods of sleep and light exposure vary widely from day to day, showing no consistent pattern. Sleep times aren't synchronized with dark times. (In addition, this person is probably light-deprived.) If your chart shows widely fluctuating patterns of sleep-wakefulness and light-dark exposure, you're likely to experience an array of confusing symptoms: sleepiness at odd times, bouts of insomnia, increased susceptibility to colds and infection, and so on. These symptoms won't necessarily track immediately with your activity levels. For example, you may feel tired three or four days after a late night. Or instead of feeling tired, you may not be able to get to sleep at bedtime. These reactions happen because melatonin production can't keep in sync with the changes in your habits.

Looking at natural rhythms. If you've also charted sunrise and sunset times, look at how they compare with your daily exposure to light and dark. Ideally, they should be in sync; to the extent they differ, you'll find it harder to keep your body's "clock" set to your schedule; it may tend to drift toward a normal daylight rhythm. That can be a problem, for example, in shift work. If you work the night shift, it's important to keep your visual cues consistent. For example, step out into the daylight for a few minutes shortly after you wake up, even if it's in the evening. Conversely, if you watch the sun come up before turning in, you may have trouble sleeping. Instead, try to get into bed before the sky gets light.

Step 2: Taking Stock

To find out how specific lifestyle factors are affecting your circadian rhythms, complete this checklist to determine whether your lifestyle is in sync with natural rhythms of light and dark. In the discussion that follows this assessment, we offer some strategies for simple changes to bring your life at home and work into better sync with your body's natural rhythms.

Living with Your Natural Rhythms

For each of the following groups of statements, choose the one that best describes your lifestyle. Then use the scoring guide to find out how melatonin-friendly your lifestyle is.

AT WORK

1. Type

_____ A. My job follows a regular schedule and the same type of things happen at about the same time each day. I always arrive at the same time and take my lunch almost always at the same hour.

_____ B. Different things happen from day to day on my job, but I'm usually able to plan for them ahead of time. Once my schedule for the week is set, it doesn't usually change much.

_____ C. My job doesn't have a set routine; when I come to work every day, I never know what's going to happen. One day I may have a meeting at 8:00 A.M.; the next day I may have to work until 10:00 P.M.

2. Travel

_____ A. I'm rarely required to take business trips that cross three or more time zones.

_____ B. A few times a year, I take a trip that crosses three or more time zones.

_____ C. More than once a month, I take such trips.

3. Shifts

_____ A. I usually work the day shift (7 to 3, 8 to 4, 9 to 5, etc.).

_____ B. I usually work the evening shift (3 to 11, etc.).

_____ C. I usually work the night shift (11 to 7, etc.).

_____ D. I don't have a regular work schedule.

4. Shift Length

_____ A. I usually work eight-hour shifts or shorter.
_____ B. I usually work ten-hour shifts.
_____ C. I usually work shifts longer than ten hours.

5. Routines

_____ A. I work regular office hours, or I work the same shift for at least six months at a time.
_____ B. I change shifts every six months or sooner.
_____ C. I change shifts once a month or sooner.
_____ D. I don't have a set routine; I never know whether I'll be working days, evenings, or nights.

6. Rhythms

_____ A. My work week follows a seven-day schedule (for example, five days on, two days off; four days on, three days off).
_____ B. My work week follows a schedule other than seven days (for example, ten days on, four days off; six days on, three days off).

7. Overtime

_____ A. I don't usually work overtime.
_____ B. I work overtime on a more-or-less regular schedule (for example, an extra hour every evening, three hours on Saturdays, etc.).
_____ C. I work overtime a lot; the hours vary, and it often comes up unexpectedly at the last minute.

AT HOME

8. Routines

_____ A. The routines around my house—getting up, eating, going to bed—are usually regularly scheduled.

_____ B. Something unexpected is always coming up at my house making it impossible to follow a regular schedule for meals and sleeping.

9. Changes

_____ A. There haven't been many changes in my life recently.

_____ B. Recent changes in my job or personal life have dramatically altered my schedules (for example, a new job, getting married, going to college, a new baby, a new puppy).

BEDTIME

10. Time

_____ A. I usually go to bed about the same time every night.

_____ B. My bedtime varies from night to night.

11. Routines

_____ A. I usually have a set routine at bedtime (for example, watching the evening news, reading for half an hour, etc.).

_____ B. I don't have a set routine.

12. Activity Level

_____ A. In the hour before bedtime, I'm usually winding down.

_____ B. In the hour before bedtime, I'm pretty active (for example, doing last-minute chores, rushing to clean up, etc.).

13. Light

_____ A. I usually sleep in total or near-total darkness.

____ B. There's usually some light when I'm sleeping (for example, a desk lamp or television).

____ C. I usually sleep in daylight (for example, during the day with the blinds open).

SLEEP PATTERNS

14. Sleeping In

____ A. On my days off, I usually get up about the same time as on workdays.

____ B. On my days off, I usually "sleep in."

15. Quantity

____ A. I feel that I get about the right amount of sleep.

____ B. I feel that I don't get enough sleep.

____ C. I feel that I sleep too much.

16. Type

____ A. I'm a sound sleeper: I usually fall asleep promptly and wake up rested.

____ B. I have trouble falling asleep, but once I do I usually sleep well.

____ C. I usually toss and turn all night; I rarely get a good night's sleep.

____ D. I'm often awake at night and sleepy in the daytime.

17. Naps

____ A. I rarely or never take naps.

____ B. I often take naps during the day and wake up feeling refreshed.

____ C. I often take naps during the day and wake up feeling groggy and disoriented.

18. History

_____ A. Basically, my sleep patterns are about what they've always been.

_____ B. My sleep patterns have changed in the past two years—I'm a lighter/heavier sleeper than I used to be.

19. Seasonal

I usually sleep better during the

_____ A. spring

_____ B. summer

_____ C. fall

_____ D. winter

MOODS AND ALERTNESS

20. General

_____ A. When I'm awake, I usually feel rested and alert.

_____ B. When I'm awake, I usually feel tired and sleepy.

21. Seasonal

I am usually most alert and productive in the

_____ A. spring

_____ B. summer

_____ C. fall

_____ D. winter

22. Daily

_____ A. I think of myself as a morning person.

_____ B. I think of myself as a night person.

23. Light and Dark

_____ A. My moods are about the same when it's bright or gloomy.

_____ B. My moods seem to shift with the intensity of light.
_____ C. Cloudy days make me irritable and depressed.

24. Phase Shifts

_____ A. I usually bounce back quickly from shift changes, daylight savings time, jet lag, etc.
_____ B. I have a lot of difficulty coping with these changes.

SUNLIGHT

25. Patterns of Exposure

_____ A. I spend most of my day outdoors.
_____ B. I work indoors but near a window, or I spend a lot of time in the car.
_____ C. In some seasons I'm outside a lot; in other seasons I usually work indoors.
_____ D. Year-round, I spend most of my day indoors under artificial light.

LEISURE AND EXERCISE

26. Patterns of Activity

_____ A. I'm usually home in bed before the 11 o'clock news.
_____ B. I usually spend quiet evenings at home, but I always catch the late-night talk shows.
_____ C. I go out at night once a week or less.
_____ D. I stay out late every night.

27. Exercise

_____ A. Year-round, I usually exercise outside, not indoors.
_____ B. I tend to mix indoor and outdoor exercise.
_____ C. I exercise outside when the weather's warm and inside when it's cold.
_____ D. Year-round, I usually exercise in a gym or at home, rarely outdoors.

28. Leisure

_____ A. Most of my leisure activities take place outdoors.
_____ B. Year-round, I tend to split my time between indoor and outdoor leisure activities.
_____ C. I tend to be more active outdoors in certain seasons.
_____ D. Most of my leisure activities take place indoors.

Scoring. *For each of the questions above, assign yourself points as follows for your responses. Total your points, then match your total score to the key that follows.*

1. Type
   A = 4
   B = 2
   C = 0

2. Travel
   A = 4
   B = 2
   C = 0

3. Shifts
   A = 4
   B = 3
   C = 2
   D = 0

4. Shift Length
   A = 4
   B = 2
   C = 0

5. Routines
   A = 4
   B = 2
   C = 0
   D = 0

6. Rhythms
   A = 4
   B = 1

7. Overtime
   A = 4
   B = 3
   C = 1

8. Routines
   A = 4
   B = 0

9. Changes
   A = 4
   B = 0

10. Time
    A = 4
    B = 0

11. Routines
    A = 4
    B = 2

12. Activity Level
    A = 4
    B = 2

13. Light
    A = 4
    B = 2
    C = 0

14. Sleeping In
    A = 4
    B = 0

15. Quantity
    A = 4
    B = 0
    C = 0

16. Type
    A = 4
    B = 3
    C = 2
    D = 0

17. Naps
    A = 4
    B = 2
    C = 0

18. History
    A = 4
    B = 1

19. Seasonal
    A = 2
    B = 0
    C = 2
    D = 4

20. General
    A = 4
    B = 0

21. Seasonal
    A = 2
    B = 4
    C = 2
    D = 2

22. Daily
    A = 4
    B = 0

23. Light and Dark
    A = 4
    B = 0
    C = 0

24. Phase Shifts
    A = 4
    B = 0

25. Patterns of Exposure
    A = 4
    B = 3
    C = 2
    D = 0

26. Patterns of Activity
    A = 4
    B = 3
    C = 2
    D = 1

27. Exercise
    A = 4
    B = 2
    C = 1
    D = 0

28. Leisure
    A = 4
    B = 2
    C = 1
    D = 0

What the scores mean. Your total score can give you a general idea of how well your current lifestyle supports your body's production of melatonin.

Higher scores suggest that you get plenty of opportunities for your body to use daylight to keep the pineal gland synchronized, and that you tend to follow well-established schedules in your daily life. Both exposure to daylight and regular schedules are important to keeping the pineal gland on track; while the basic rhythm is set by light, pineal research indicates that the body also uses other "cues" from the environment to keep itself regular—everything from bedtime routines to your morning shower.

In addition to your total score, also look at your individual responses. For any individual question, a score of 3 or 4 indicates a good match between your lifestyle and natural rhythms. Scores of 0 to 2 suggest potential problem areas.

40 or below. Your routines tend to change often and you're not getting consistent exposure to natural daylight. Your body's production of melatonin is often out of sync with your daily routines, leaving you prone to fatigue, illness, and mood swings. See the section that follows for suggestions on how to make your lifestyle more melatonin-friendly.

41 to 79. Your lifestyle is somewhat synchronized with the sun, but there are ways to make it more so. You will also benefit from the suggestions that follow.

80 or above. Congratulations. Whether by accident or design, your lifestyle is closely synchronized with natural patterns of light and dark.

Strategies for a Solar-Centered Lifestyle

Armed with the information you've collected, you can now look at each of the specific areas identified in the lifestyle inventory. In this section, we offer some strategies and suggestions that can help you keep your body's clock in sync and optimize your production of melatonin.

Your goal is simple in theory, though it may get complicated in real life: *To duplicate, as much as possible, the daily rhythms of light and darkness that our preindustrial ancestors were exposed to.* Here's a rule of thumb: Live in the light, sleep in the dark. Ideally, we should seek to get exposure every day to sunlight beginning as soon as possible after we get up, and go to sleep in the darkness, after our body begins to produce its evening supply of melatonin.

What about artificial light? Ordinary artificial light isn't sunlight—and it can't fool your body. The lights in your office and home may seem bright, but they're far less intense than daylight, even on dim winter days. Full-spectrum high-intensity fluorescent lights, in special reflective cabinets, used to treat seasonal affective disorder (SAD), do mimic the sun's effects, but short of investing several hundred dollars to install them in your home, you'll have to rely on the sun.

What about cloudy days? Most cloudy days are bright enough to keep your body's clock in sync. And even an occasional dark day isn't enough to upset your basic circadian rhythm, although you may notice that you feel sleepy on really gloomy days. People with SAD often notice a worsening of symptoms after a few cloudy days in a row. But for most people, it's the long-term trends that matter, and your internal clock can keep itself on track, especially if other routines continue to follow the same schedule.

At Work

Type of work. The type of work you do has a big impact on your body's ability to maintain its natural rhythms. If your job follows predictable routines and schedules, it's easier to keep your body on track. If you work during the day and sleep at night, that's even better. If, on the other hand, you have the kind of job that often involves late nights, "crunches," and last-minute meetings called for 6 A.M., you're at greater risk for upsetting your body's natural melatonin production. While long hours per se aren't a problem, *unpredictability* of your schedule is.

Most times, there's not too much you can do about the basic nature of your job, and many kinds of work aren't amenable to set routines. But sometimes you can plan or reschedule activities to minimize the effects of erratic work schedules. For example, if you must attend early-morning meetings, try to leave some breathing room before the rest of your day starts for a walk around the block. These few minutes of exposure to sunlight helps keep your internal clock synchronized even when your day starts before dawn.

Pacing. Try to pace your work so that you keep to a more-or-less regular schedule. For example, it's better to put in one extra hour a day rather than saving it all up and then working until eleven o'clock one evening. If you have a habit of working several long nights and then crashing afterward, you may find that your body can't keep up. Some variation in your workday is normal and healthy, of course. But as a general rule of thumb, try not to vary your work hours by more than 20 percent from day to day.

Taking advantage of natural rhythms. By understanding how melatonin regulates your body's clock, you can also structure your activities. As melatonin levels begin to rise in the evening, you can expect to feel lethargic, even if you slept late in the morning. Knowing this, it may be best to schedule the most intensive work earlier in the day, and use the evening to catch up on less demanding tasks. At the very least, you will know to allow extra time to complete work in the evening.

Travel. If your business travel takes you to different time zones, you're probably putting chronic stress on your pineal gland. If possible, consolidate your trips: A single two-week trip to another time zone upsets your time clock less than seven two-day trips. If you travel on multiple-city trips, try to arrange your itinerary to make it melatonin-friendly by visiting all the cities in one time zone before moving to the next one. If you must travel across time zones, consider taking melatonin supplements to reduce the effects of jet lag (see page 190).

Shift work. One of the greatest stressors to your body's rhythm is shift work, especially if you work double shifts, erratic shifts, or short rotations. Shift work keeps your body in an almost constant state of "jet lag," as your body's natural clock struggles to keep up. Of course, you usually can't do much to change your actual shift schedule, but here are some strategies to help minimize the adverse effects:

First, encourage your employer to consider longer rotations. If you change shifts once a season rather than once a month, your body has to adjust to only a third as many changes every year—and you'll still work the same number of days, evenings, and nights over the course of the year.

Also, try to keep your personal clock on the same schedule as your professional one. If you work the night shift but operate on "day shift" hours on your days off, you don't give your body's rhythms a chance to reach a balance. Granted, it's tough to structure your days off to fit a night-shift schedule—to find an open grocery store or bank, or to plan family activities—but it may be possible to strike a balance. For example, try to schedule more afternoon and evening activities on your days off, and sleep later in the morning. If your shifts *do* change often or unpredictably, you may be better off keeping your internal clock synchronized with daylight. For example, try to get some sunlight in the morning, and try to get some sleep at night.

A more radical approach to dealing with shift changes is to consider the option of a permanent shift. Many hospitals, for example, pay a differential to nurses who are willing to work the night shift permanently. While this option isn't for everyone, from the standpoint of your internal clock it's easier to adapt to a permanent night shift than to frequent shift changes.

When you do work the night shift, use the sun to your advantage. By controlling your exposure to sunlight, you can to some degree mimic the effects of normal daylight—in effect, "fooling" your pineal gland into thinking day is night and vice versa. As a general rule of thumb, keep in mind that *darkness makes you sleepy and light makes you wakeful*. So no matter what your schedule, avoid sunlight before going to bed and seek it out when you wake up.

For example, after you get off work and before you go to bed, stay out of the sun if you can; otherwise, the sunlight will cue your body to stop producing melatonin and you may find yourself restless and unable to fall asleep. Sleep in a darkened room, and when you wake up, get into the sun as soon as you can: The pineal will "think" it's morning and shut down its melatonin production. You may actually feel more rested if you set your alarm to wake up an hour before sunset, rather than sleeping later and getting up in the dark. (As with jet lag, melatonin supplements can help ease the transition to new shifts; see chapter 15.)

Weekly rhythms. Some people work shift rotations that aren't in sync with weekly schedules. There's nothing inherently magical about a seven-day week, but these kinds of rotations can result in added stress by causing conflicts between work and home schedules, adding to their unpredictability. Remember, *changes* in schedules tend to throw off your internal clock.

At Home

Routines and changes. Regular schedules and routines help us to keep our clocks synchronized and avoid the feelings of disorientation and fatigue that occur when our patterns of activity are at odds with melatonin production. Indeed, erratic schedules can cause symptoms similar to those of jet lag, and for similar reasons: We are active at times that our bodies are physically geared up for sleep, or trying to sleep when our bodies are in a wakeful mode. The new baby or the puppy that's being housebroken, changes in living arrangements, or even moving to a noisier apartment upsets our natural rhythms.

Of course, some degree of change is normal and healthy. But look at whether you've been asking your body to accept too much unpredictability, and consider ways to manage change: If you've just moved to a new house, for example, give yourself a few months to settle in before you get that puppy. Or use the VCR to tape that late-night movie and watch it the next evening.

Bedtime

It turns out Mom was right all along: A regular bedtime is important. Not necessarily early, but regular. If you're the type who tends to stay up into the wee hours one night and then turn in after dinner the next, it's hard for your body's clock to stay set. The effects may not be immediately apparent, but a few days later you may find yourself pacing the halls in the middle of the night or dozing at your desk before lunch.

Mom was right about those bedtime rituals, too. They help us sleep. Light and darkness are the body's primary timekeepers, but research suggests that the body uses other means to help keep itself in sync as well. We associate certain behaviors—certain rituals—with periods of sleep and wakefulness, and these rituals can help sustain circadian rhythms and regulate melatonin production even when visual cues are absent or misleading, such as when people are confined indoors or the skies are overcast for several days. So when you follow those rituals that help you sleep—checking the doors even though you know they're locked, drinking a glass of milk, fluffing the pillow—you're being sensible, not silly. (That glass of milk, by the way, helps in another way as well: It contains tryptophan, one of the building blocks of serotonin and melatonin.)

Your level of activity before bedtime can give you some insights into the functioning of your body's circadian rhythms and, indirectly, its production of melatonin. Normally, as bedtime approaches and your body begins to produce melatonin, you start to feel sleepy and your activity level decreases. If you feel wide awake right up until the time you climb into bed, it *may* be a signal that your body's melatonin production is out of sync with your schedule. (Similarly, feeling drowsy during the day may be an indication that your sleep cycles don't match your melatonin cycles.)

A number of other factors can mask or override the effects of melatonin, for example, drinking coffee or alcohol in the evening, or experiencing anxiety or tension. But if you don't feel sleepy when you should and there's no apparent reason, take a close look at your schedule and its relationship to light and dark.

Sleeping in the light. Current studies suggest that you need daylight or powerful artificial light to influence the pineal gland; low-level artificial light doesn't seem to register. In short, there's no evidence to suggest that sleeping with a light on will interfere with normal pineal activity.

However, we also know that the pineal can be affected in more subtle ways by a range of influences. For example, low-energy EMFs affect the pineal gland, even in the absence of any light at all. Similarly, scientists may find one day that sleeping under artificial light may have subtle and/or long-term effects on the pineal gland or the body's natural rhythms. If you usually sleep with one or more lights on, try sleeping in darkness for a week or so and see if it improves your sleep patterns or daytime alertness.

If you work nights or evenings and sleep during the day, make sure that you go to sleep in a room that lets as little daylight in as possible. Keep the blinds closed and the curtains drawn.

Sleep Patterns

Sleeping in. For shift workers and nonshift workers alike, it may be time to rethink the time-honored tradition of "sleeping in" on your days off. Researchers haven't studied the effects of sleeping in, but it's reasonable to assume that when we sleep in more than a couple of hours, we interfere with our bodies' attempt to establish regular rhythms. In fact, it's possible that sleeping in on weekends may contribute to the lethargy of "Monday-morning blues." Rather than being a psychological letdown at the prospect of the coming week's work, these blues may be a physical reaction caused by delayed melatonin production. Instead of sleeping late on weekends, try getting up about the same time as you do on weekdays. You may find that you feel *more* rested, both on the weekend and during the week.

Sleep duration. The old myth of needing "eight hours' sleep" is just that. (Some believe it dates back to the Romans, who with their passion for order divided the day into three equal parts, one of which was for sleeping.) Many people function well on five hours a night or less; others need ten or twelve. The problems come when we get less than we need, or sleep more than we should.

The amount of sleep you should get is the amount you need to feel rested and alert during the day.

If you're sleeping too much or your sleep patterns are erratic (for example, tossing and turning at night and napping in the daytime), it's often a sign of depression or another underlying problem. In addition, many drugs (including those that seem to make you sleepy, like alcohol, Valium, or marijuana) actually interfere with normal sleep rhythms. Even though you may be sleeping more, you're getting less rest. If you're sleeping a lot but not feeling rested, see your doctor to find out if there's a medical reason.

If you're sleep-deprived, you're probably also melatonin-deprived. In addition to feeling sleepy during the day, you may also experience more colds, flu, and other infections, as well as long-range health problems.

As people get older, they find it harder to maintain a normal nightly sleep rhythm. This problem seems to be a direct effect of reduced melatonin production by the pineal gland. It's often said that people need less sleep as they get older. That's not true: They *sleep less,* but they still need the same amount of sleep, which is why they often feel the need to take catnaps during the day. If your sleep patterns have changed as you've grown older, you may be a candidate for melatonin supplements (see chapter 15). By restoring your nightly melatonin levels to those of youth, these supplements literally help you "sleep like a baby."

Naps. Naps are often (though not always) a sign that something's wrong, especially if you only recently started to feel the need for them and there isn't any obvious reason (such as a recent shift change). As you begin to establish better sleep patterns, the need for naps should diminish.

Changes in sleep patterns. Another early warning sign of melatonin depletion is a shift in long-established sleeping patterns for no apparent reason. Lots of things, large and small, can affect our ability to sleep, and they're not always obvious. A new job, for example, may establish different rhythms to which your body must adjust. But if your sleep habits change dramatically and you can't figure out why, see your doctor. There may be a physical reason—anything from a developing ulcer to heart disease.

Seasonal variations. Some variation of sleep patterns across seasons is normal and is related to shifting patterns of light and dark; most of us tend to sleep more in the winter months and less in the summer. When these shifts are unusually large, they may be a sign of SAD (see chapter 16), especially if they're accompanied by mood swings. For everyone, it's best to keep in mind the relationship of the seasons and sleep, and to plan accordingly. (Ironically, the season when we tend to feel sleepiest—the dark days of December—also tends to be the one with extra demands on our time. The greater incidence of mental illness and suicide around the holidays may be related to this unfortunate combination of dark days and long hours.)

Light-Related Mood Swings

Remember the lyric "Rainy days and Mondays always get me down"? The reason may be related to melatonin. There's growing evidence that melatonin has powerful effects on mood. As we saw previously, shifts in melatonin production caused by "sleeping in" might help explain the Monday-morning blues. Similarly, the lack of sunlight on rainy days could be why our moods are often as gloomy as the skies. These effects are generally temporary and harmless; however, if you experience *severe* mood changes related to light and dark, ask your doctor to evaluate you for SAD.

It probably doesn't matter too much whether you're a night person or a day person, in terms of when you're most active and alert. However, if you *are* a night person, you may be at greater risk for melatonin disruptions, simply because you're likely to get less exposure to daylight. If your best time is nighttime, you may need to make a special effort to get lots of daylight exposure, especially in the morning.

Phase shifts. In the terminology of pineal research, "phase shifts" occur when the external world and the body's internal clock are quickly taken out of sync. In the real world, you experience this kind of shift when, for example, you fly across the continent, change shifts at work, or (to a lesser degree) reset your alarm for daylight savings time. When your body "thinks" it's noon, the world around you is operating at 3 P.M., 5 A.M., or 11 A.M. The fatigue and disorientation you feel—jet lag—occurs as your body resets its clock to fit the surroundings.

Some people have naturally resilient internal clocks that swiftly adjust to phase shifts; others require a longer period of adjustment. If you're one of the latter, try moving through phase shifts gradually. If you live on the East Coast and you've planned a vacation in Hawaii, for example, try setting your alarm an hour earlier two weeks before you leave, two hours earlier a week before you leave, and three hours earlier for the last two days before you leave. Use a similar strategy for shift changes. The shift to and from daylight savings time isn't usually enough to throw off our internal clocks for more than a day or two, but if it's tough for you, consider moving your alarm forward or backward fifteen minutes a day in the four days preceding the change.

Sunlight

If you have the kind of lifestyle that keeps you exposed to lots of daylight, congratulations. Farmers, mail carriers, roofers, and lifeguards don't have to worry too much about managing their circadian rhythms—the sun does all the work.

People who drive a lot—delivery people, truckers, sales representatives, and moms, for example—also get a lot of exposure to daylight. So do office workers who work near large, unshaded windows. However, it's not yet clear from the research whether glass windows and windshields block the effects of daylight on the retina and pineal gland. Ordinary glass blocks the transmission of invisible ultraviolet (UV) light—the part of sunlight that causes both sunburns and suntan. It's possible that these UV rays are needed to help maintain circadian rhythms. But since people who drive a lot tend to be in and out of their cars during the day, it's likely that their melatonin production is well regulated anyway.

If you spend most of your time working indoors, your body's exposure to daylight may be less than ideal. The deficiency is likely to be greater during the late fall and early winter, when the nights are long. (In northern cities in the month of December, it's not unusual for an office worker to receive only a few minutes of exposure to daylight a day.) To maintain a vigorous cycle of serotonin and melatonin production, make an effort to increase your exposure to daylight, especially when the days are short. Take a walk at lunchtime rather than staying at your desk or in the company cafeteria. If you can, position your desk near a window. On weekends, try to get as much exposure to the sun as possible, especially early in the morning. If you have flextime, you may be able to adjust your work hours to get more exposure to daylight before or after the workday.

Leisure and Exercise

Your leisure time affords additional opportunities to keep your inner clock synchronized with the sun. For example, you can design an exercise program that helps keep your pineal gland as well as your muscles in good shape. If you exercise exclusively at a gym, consider moving some of your activities out of doors: Substitute a lunchtime or early-morning run through the neighborhood for a session on the treadmill, for example. Or use an outdoor tennis court rather than an indoor one.

Also keep in mind your body's natural rhythms of activity when you exercise. As melatonin production begins in the evening, you may find you have less energy for your exercise routine. If you've been trying to follow a regular exercise program but haven't been able to stick to it, you may have better luck if you switch to a morning workout. Further, strenuous evening exercise may interfere with melatonin's sleep-inducing effects and make it hard for you to get a good night's sleep.

In other leisure activities, look for ways to enhance your exposure to daylight. Remember, the pineal gland is regulated by the eyes, so full body exposure isn't necessary. You don't need to go sunbathing on the beach; all you need is to spend time in natural daylight, even if you're covered from head to toe in a ski suit.

Exposure to the sun is especially critical in the wintertime, when days are short. Ironically, these are the times when we're most likely to replace outdoor activities with indoor ones. Whether your preferences run to cross-country skiing or landscape painting, look for winter activities that get you into the light.

Fine-Tuning the Biological Clock

Periods of light and dark actually fine-tune an underlying rhythm that's built into our biological system. Left to itself, this rhythm runs on a cycle of approximately twenty-five hours; the light-dark cycle adjusts the rhythm ("entrains" it, in the language of the researchers) to a twenty-four-hour rhythm. Nobody knows for sure why the underlying rhythm runs an hour longer than a day.

The pineal gland has some help keeping the body's rhythms in sync. Behavioral factors assist in the "entraining" of circadian rhythms. Events that occur at certain times during the day—meals, bedtime, work hours, and so on—all provide cues that help keep the body on schedule. Though their effects aren't as powerful as the light-dark cycle, they reinforce it and help carry the rhythms through periods (such as dark rainy days) when the light rhythms aren't as pronounced. Scientists call such cues *zeitgebers*—German for, literally, "time-givers." Working together, zeitgebers keep the biological clock from drifting off center.

Light-to-dark differences of 3,000 lux or more (the difference between, say, daylight and bright incandescent light) are the strongest zeitgebers. They have a "range of entrainment" of about plus or minus four hours. In other words, they can move the circadian clock forward or backward about four hours at a time. If the time shift is greater than this, the changing light-dark patterns won't entrain your body's rhythms. So, for example, if you fly from the East Coast to the West, you can use daylight to reset your clock; if you fly to Korea, it won't work. Other zeitgebers are less effective.

After light-dark cycles, the second strongest zeitgebers are patterns of behavior; they can entrain rhythms plus or minus about two hours. Research shows that experimentally induced changes in magnetic field strengths have an entrainment range of about one hour. Next are light-dark shifts involving artificial light (difference of 300–1,000 lux) and temperature differences (warm/cold).

In the absence of any zeitgebers, the natural rhythm runs on a cycle of about twenty-five hours. We don't know why. In experiments where people live for weeks or months without any time cues, they tend to get up about an hour later each day. After two weeks, they're getting up in the middle of the night and sleeping during the day. After about a month, they're back on the original schedule.

It turns out that similar patterns are often seen in people with autism, Alzheimer's disease, blindness, and other conditions in which circadian rhythms are disturbed. These people essentially operate on a twenty-five-hour day while the rest of us are on twenty-four-hour days, which explains, in part, the propensity to night roaming and severe sleep disturbances. In a similar vein, newborn infants often lack the biological hardware to adjust to a twenty-four-hour schedule. As their nervous systems mature, they develop an increasing ability to "track" onto a daily schedule.

Thus, our knowledge of zeitgebers can give us some practical insights into helping people with disturbed circadian rhythms and the people who care for them. Already researchers are exploring the use of melatonin in Alzheimer's disease and autism. At the same time, caregivers and families should reinforce natural "entrainers" as much as possible: exposure to daylight, behavioral cues (for example, serving dinner at the same time every day, the use of bedtime rituals), even low-light exposure (ensuring the person sleeps in a darkened room).

For new parents, similar measures might help your baby establish a good sleep schedule sooner. Although to a certain extent you simply have to wait until your baby's neurological system matures, it's a good idea to make these same "entrainers" available to him or her as soon as possible: daily exposure to sunlight, especially in the morning (if need be, through the window in the wintertime); consistent behavioral cues (for example, put him or her to bed every night in the same crib, at the same time, positioned in the same direction); and, to the extent possible, avoiding artificial light in the nursery at night.

The strategies in this chapter will help you bring your life's rhythms into harmony with those of the natural environment. Unlike many other health regimens, most of these changes are easy to make and easy to sustain.

Though they may seem minor, they can pay big dividends now and in the future. As you move your lifestyle more closely into alignment with your body's natural phases of melatonin production, you will likely find that you sleep better at night and have more energy during the day. You may also discover that you have fewer colds and infections, more stable moods, a greater sense of well-being and alertness, and increased productivity. And you're also helping protect your body's organs and tissues from damage and keeping them working well throughout your life. As the old proverb says, "Early to bed and early to rise, makes a man healthy, wealthy, and wise."

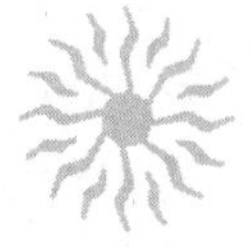

CHAPTER 10

The Melatonin-Friendly Diet

In this chapter, we'll look at some ways you can enhance your body's supply of melatonin through diet, without radically changing your current eating habits. These strategies don't require a lot of sacrifice; they're meant to allow you to make adjustments in the way you eat in light of what we know about melatonin and how it's produced.

If you're now on a special diet—to lose weight, to reduce cholesterol, to restrict intake of sugar, or for any other reason—you can and should stick to your program. But within the guidelines of that diet, consider some of the following changes to help your body increase its melatonin output.

Boost Your Antioxidants

That old television commercial was right: Breakfast without orange juice *is* like a day without sunshine. Orange juice and sunlight both enhance your ability to fight free radicals.

As we saw in chapter 4, melatonin has some unique properties that make it the most effective free-radical scavenger known. But you can give it some added help by making sure your diet is full of other free-radical-fighting substances. Also known as antioxidants (because they block oxidation, the chemical reaction caused by free radicals), these substances augment melatonin's action. And every free radical that they eliminate frees up more melatonin to fight elsewhere.

Vitamin E

After melatonin, the strongest antioxidant known is vitamin E. Researchers have found, for example, that vitamin E is effective in preventing free-radical damage from strenuous exercise. Studies have also shown that vitamin E supplements reduced rates of prostate cancer and colorectal cancer. Male smokers who took vitamin E had 5 percent fewer deaths from heart disease. Numerous other studies show similar effects and confirm vitamin E's powerful antioxidant properties. You can increase your intake of vitamin E by eating foods that are naturally high in this substance.

Food Sources of Vitamin E

Beets	Grains
Black beans	Mangoes
Brown rice	Nuts
Cold-pressed polyunsaturated vegetable oils (corn oil, safflower oil)	Peanut butter
Dark green leafy vegetables (spinach, kale, broccoli)	Sunflower and other seeds
	Sweet potatoes
	Wheat germ

Vitamin C

Vitamin C is another powerful antioxidant, found not only in orange juice but in other citrus, fresh fruits, and vegetables. Fifteen separate studies support a link between vitamin C and reduced risk of stomach and esophageal cancer. Other studies tie vitamin C to reduced risks of cancer of the lung, larynx, colon, rectum, pancreas, cervix, and bladder. And a study at Tufts University in Boston found lower blood pressure readings among people whose diets were rich in vitamin C.

Our bodies neither store nor manufacture vitamin C; it must be replenished continuously. There is some controversy over how much vitamin C we need every day. The official U.S. Recommended Daily Allowance, which has ranged from 45 to 60 milligrams, was established long before we understood the role of vitamin C in preventing free radical damage. It was set at a level to prevent scurvy—a disease caused by severe deficiencies of vitamin C—and not to provide antioxidant protection. A more realistic daily level is 2,000 to 5,000 milligrams.

Food Sources of Vitamin C

Canteloupes	Oranges, orange juice, other citrus
Elderberries	Papayas
Guavas	Strawberries
Kiwis	Tomatoes, tomato juice
Mangoes	

Beta-Carotene

Beta-carotene, the substance that makes carrots orange and tomatoes red, is also a potent antioxidant. In a Canadian study, seniors taking a daily multivitamin, supplemented with extra doses of beta-carotene and vitamin E, were less likely to get sick. And when they did get sick, they got well faster than a comparable group not taking vitamins. Other studies have shown beta-carotene stimulates the immune system, helps prevent ultraviolet radiation from damaging immune functions, and produces potentially protective effects against lung cancer.

Beta-carotene also offers protection against heart disease. A Harvard study found that women who ate the equivalent of one large carrot a day had a 22 percent lower risk of heart attack and 40 percent lower risk of stroke than those who didn't. In another study, a dose equivalent to about three carrots a day cut the risk of heart attack and stroke almost in half.

(A Finnish study, however, raises a caution about beta-carotene. In a group of 29,000 middle-aged smokers, it appeared that people who took 20 milligrams of beta-carotene supplements a day for five to eight years apparently didn't gain any protection from lung cancer;

in fact, the group that took beta-carotene had a slightly *higher* risk for lung cancer. However, there were serious shortcomings in the methodology of this study, and there is a great deal of other evidence demonstrating beta-carotene's health benefits. Scientists are looking at these findings carefully, and in the meantime, the opinion is nearly universal that a diet high in beta-carotene is good for you.)

To increase beta-carotene in your diet, look for yellow and orange fruits and vegetables such as canteloupes, yams, sweet potatoes, and carrots. Leafy green vegetables such as spinach, kale, and broccoli are also high in beta-carotene (the green chlorophyll masks the orange color).

Cruciferous vegetables—those in the mustard family—including brussels sprouts, broccoli, cabbage, kohlrabi, bok choy, collard greens, mustard greens, cauliflower, kale, rutabagas, and turnips, are another good source of beta-carotene. In addition, these vegetables contain *indoles*—chemicals related to melatonin with potentially cancer-fighting properties.

Food Sources of Beta-Carotene

Apricots
Beet greens
Bok choy
Broccoli
Brussels sprouts
Cabbage
Canteloupes
Carambola (star fruit)
Carrots
Cauliflower
Collard greens
Guavas
Kale
Kohlrabi
Mangoes
Mustard greens

- Nectarines
- Papayas
- Peaches
- Pumpkins
- Rutabagas
- Spinach
- Sweet potatoes
- Tangerines
- Turnips
- Winter squash
- Yams

Vitamin A

Related to beta-carotene, vitamin A helps fight infection, especially respiratory infection; it seems to prevent bacteria from clinging to the linings of the lungs and respiratory tract, and it stimulates the production of lymphocytes. In addition, vitamin A deficiencies have been linked to accelerated atrophying of the thymus.

As with other vitamins, the best sources of vitamin A are dietary—primarily meat, poultry, fish, eggs, and milk fat. Also, foods rich in beta-carotene have an added benefit, since the body converts it into vitamin A.

Selenium

Another key antioxidant that may also exert anticancer properties, especially against stomach and esophageal cancer, is selenium. A caution, however: Too-large doses may impair the immune response, and cause loss of hair and nails. The body appears to require only trace amounts of this mineral, and it's most easily obtained from eating grains. In my practice I usually like patients to receive approximately 200 mcg/day.

Food Sources of Selenium

- Fish and seafood
- Lean red meat
- Milk and dairy products
- Whole-grain cereals and breads

Preserving the Antioxidant Value of Your Foods

Boosting antioxidants in your diet involves not only the types of foods you eat, but how you store and cook them. Processing, storage, and cooking methods can all rob foods of their natural antioxidant content.

Vitamin E is an oil-based vitamin, and it tends to remain stable in foods. However, fruits and vegetables containing beta-carotene usually manufacture more of it as they mature—as shown by their increasingly red, yellow, or orange color. (Think of a pumpkin or canteloupe ripening on the vine.) In general, the riper the food, the more beta-carotene it will contain.

As a general rule, fresher is better when it comes to foods containing vitamin C. Vitamin C breaks down rapidly when foods are exposed to air or heat. Citrus products and sweet peppers retain their vitamin C longer than other fruits and vegetables.

Freezing. Freezing per se doesn't significantly affect the antioxidant content of foods. However, if the foods are blanched before they're frozen, they're likely to lose antioxidants—particularly vitamin C, which is water soluble.

Canning. Canning also destroys between one-fourth and three-fourths of foods' vitamin C content, mostly because of the heat used during processing. High temperatures during storage can also break down vitamin C and beta-carotene. Canned tomatoes retain more of their vitamin C than other canned foods.

Thawing. Use thawed foods promptly. They may lose a significant portion of their nutrients within the first twenty-four hours of thawing.

Cooking. Cooking destroys some antioxidants. When frozen vegetables are cooked, they lose about 30 percent of their vitamin C content and about 5 percent of beta-carotene levels.

How Much Do You Need?

It's not always clear what the "ideal" levels are for such antioxidants as vitamins C and E. Many researchers feel that the official Recommended Daily Allowances are too low to provide full benefits. These RDAs were originally developed decades ago, long before we understood the importance of antioxidants, and were established to prevent wholesale and obvious vitamin deficiencies. Today, we know more about such substances' vital long-term health benefits, but there's little consensus among researchers about appropriate daily dosages.

In general, you can consider the official RDAs as a bare minimum; higher levels confer additional benefits. On the other hand, there can be dangers from taking megadoses of vitamins if not supervised by a medical practitioner knowledgeable in nutrition. There have been a few reports in the literature of large daily doses of vitamin C leading to the formation of kidney stones in some people. I have routinely used large doses of vitamin C with no significant side effects. Vitamins A, D, and E are stored in the fat cells. Levels should be monitored at very high doses. Also, look first to diet rather than supplements; it's hard to go wrong with whole grains, fresh fruits, and vegetables.

Consider taking additional antioxidants at times when you're at greater risk for free-radical formation, for example, if you train for athletic events, if you're exposed to high levels of pollution, if you're working around pesticides or other toxic chemicals, or if you're a smoker.

Here are some shopping tips for supplements:

- Check expiration dates. Supplements lose their potency if they're too old. Also, keep in mind that you'll be taking the supplements over the course of several months.
- Remember that vitamin-fortified foods are just like supplements. Take them into account when you're figuring out how much to take.
- Use vitamin supplements as *supplements, not substitutes*. They can't make up for a poor diet. Take another look at the dietary recommendations above to boost your natural intake of antioxidants and other melatonin-friendly substances.

Build A Cancer-Fighting Diet

To augment melatonin's cancer-fighting properties, incorporate proven anticancer foods into your diet. The benefits of high fiber and low fat are well known, but the National Cancer Institute is also studying the following foods for the potential to protect against cancer:

- garlic
- umbrelliferous vegetables (a family that includes carrots, parsnips, and parsley)
- licorice root extract
- soybeans
- flaxseed (a grain that's widely used in Europe)

Other less well-known foods with potential anticancer activity include the following:

- Green tea—the unprocessed tea that's most often found in Japanese restaurants—may contain a substance that inhibits the growth of certain tumors. (English and American teas are usually processed, turning the tea leaves black.) Studies are under way in China and Japan to study green tea's anticancer properties.
- Fish oil. Studies at Boston University School of Medicine suggest that fish oil may help fend off colon cancer.
- Spices. Cumin, basil, and poppy seed may protect against stomach cancer, according to animal studies. Although the amounts used in these experiments were higher than the average cook would employ, smaller amounts may also confer some protection.

In addition, follow these cancer-fighting diet tips.

- Limit fat intake. Expert opinion varies on how low the fat content should be, although all agree that most Americans eat too much fat. Some suggest that fat should be as low as 10 percent of all calories consumed, although such a diet is difficult to follow. Reducing fat intake to 20 percent would represent a significant improvement in the diet of most people.

- Increase fiber to 20 to 35 grams a day by eating more whole-grain products (whole wheat bread, oatmeal, whole-grain cereals, and so on).
- Cut back on highly processed foods (e.g., smoked, pickled, salt-cured products). Such foods, when eaten in large quantities, have been linked with gastric and esophageal cancer.

Feed Your Immune System

Certain foods, by enhancing your immune response, can boost the ability of melatonin to help fend off infections and other immune-related diseases. Here are some immune-enhancing vitamins and minerals to incorporate into your melatonin-friendly diet.

Vitamin B_6

Vitamin B_6, also known as pyridoxine, is essential to the proper functioning of the immune system, especially as we grow older. Studies of older adults found that when diets are low in B_6, the immune system doesn't manufacture as many lymphocytes or as much interleukin-2. Some nutritionists recommend that most adults should get 2 milligrams a day of vitamin B_6 from diet or supplements, and at least 3 milligrams a day for older people. You can obtain 2 milligrams a day of vitamin B_6 by eating one large banana, 6 ounces of chicken breast and a baked potato. Older people would have to add 6 ounces of broiled salmon to reach 3 milligrams daily. Optimal doses are much higher, 25 to 100 milligrams a day. To obtain this amount, supplementation is needed. There have been reports of nerve damage with very high doses (above 2,000 milligrams per day).

Food Sources of Vitamin B_6

Avocados	Poultry
Bananas	Salmon
Beans	Soybeans
Lean beef and pork	Tuna
Nuts	Wheat germ
Potatoes (leave the skin on)	Yeast

Zinc

Zinc is essential for healing of wounds and sores, and helps the immune system recognize and repel foreign invaders. Even mild zinc deficiencies can impair the immune system by 20 to 30 percent. Recommended Daily Allowance for zinc is about 30 milligrams. Two raw oysters will supply more than twice the daily allowance for zinc. Or you could have an ounce of wheat germ for breakfast, then shrimp cocktail and a shoulder lamb chop for dinner to obtain the zinc you need each day. People who do not eat meat should probably consider a supplement.

Food Sources of Zinc

Fish	Oysters
Green vegetables	Poultry
Lean red meat	Whole-grain cereals and breads
Legumes (beans and peas)	
Nuts	

Copper

Though copper has not been studied extensively, early results suggest that it too has a big effect on immunity. There's no official Recommended Daily Allowance; nutritionists recommend from 2 to 3 milligrams up to a maximum of 5 milligrams per day. It is difficult to get enough copper from food alone and supplementation is recommended. Deficiencies in copper have been shown to impair resistance to disease and other immune functions.

Food Sources of Copper

Fish and seafood	Nuts
Green vegetables	Poultry
Lean red meat	Whole-grain cereals and breads
Legumes	

Finally, limit your consumption of alcohol, fat, and cholesterol to keep your immune system humming. They tend to blunt the immune response and leave you more prone to infection.

By ensuring that you get enough of these "melatonin-friendly" vitamins and minerals, you allow your body's own natural antioxidant—melatonin—to perform its important tasks. The following are guidelines I offer my patients:

Optimal Dosages for Key Antioxidants and Nutrients

Beta carotene	15,000 I.U.
Bioflavinoids	250–1,000 mg
Copper	2–3 mg
Folic acid	400–800 mcg
Selenium	200 mcg
Vitamin A	10,000 I.U.
Vitamin B_6	50–100 mg
Vitamin B_{12}	500–1,000 mg
Vitamin C	2,000–4,000 mg
Vitamin D	400 I.U.
Vitamin E	400–600 I.U.
Zinc	30 mg

Increase Your Tryptophan

Eating foods that contain tryptophan may enhance your body's ability to manufacture melatonin. Such foods include dairy products, poultry, eggs, and tuna.

The body uses tryptophan to make serotonin, which, in turn, is transformed into melatonin. Your body can't make serotonin without it, and tests have shown that low-tryptophan diets lead to serotonin deficiencies. In fact, in one reported study, a tryptophan-restricted diet worsened symptoms of depression, presumably by affecting serotonin levels.

The digestive system also seems to convert dietary tryptophan into melatonin. However, be warned that we don't yet know how dietary tryptophan affects *relative* levels of serotonin and melatonin. It could increase both—and as we've seen, high serotonin levels tend to undo the benefits we derive from melatonin. Further research will be needed before we can understand this relationship fully.

In the meantime, however, many people do find that tryptophan in the diet makes them feel better, and there's some indication that it helps enhance melatonin rhythms. For example, many find that foods high in tryptophan are natural sleep promoters. Turkey, for example, is high in tryptophan, which may account in part for the popularity of the traditional after-dinner nap on Thanksgiving.

Several years ago, tryptophan supplements (that is, tablets containing tryptophan) were taken off the market after several deaths occurred. However, the deaths were traced to contaminants in the supplements; naturally-occurring tryptophan itself isn't hazardous, nor are foods containing it.

Food Sources of Tryptophan

Cheese	Poultry (especially turkey)
Eggs	Tuna
Milk	

Limit Your Calories

Remember those experiments with the feel-young mice, in which dietary restrictions helped them live longer? We now know that the results were probably due to increased melatonin. The evidence suggests that *reducing calories in the diet stimulates greater melatonin production in the digestive tract.*

Of course, to get the same degree of life extension as the mice in the experiments, you'd have to begin a near-starvation diet in childhood, and keep it up all your life—an approach that's neither practical nor desirable. But a reasonable approach to reducing your calories can augment your other melatonin-enhancing strategies and make a big difference in the amount of melatonin your body manufactures.

Note that it is the intake of *calories*—not fat or cholesterol—that makes the difference in melatonin production. There are, of course, many good health reasons for reducing fat and cholesterol, but they don't seem to affect melatonin levels.

Control Your Iron Intake

Iron is essential for the production of hemoglobin—the red pigment in the blood that carries oxygen from the lungs to the body's cells. When iron supplies are deficient, the body can't manufacture sufficient quantities of new red blood cells, resulting in anemia. Such anemic symptoms as fatigue and lack of energy reflect the body's inability to supply enough oxygen to its cells. Women are at greater risk of anemia than men, especially during pregnancy, when the developing fetus "borrows" iron supplies from the mother.

But none of this means you should be taking iron supplements routinely, and it doesn't mean that more iron will make you feel better. In fact, *iron is a powerful generator of free radicals.* So if you're taking iron supplements, be sure you really need them.

Your doctor can perform a blood test to determine if you're anemic. If you do have iron deficiencies, look to increase your dietary intake first; use supplements only if blood tests show that these dietary changes haven't reversed the anemia. If you're a vegetarian, you're at greater risk since vegetable sources of iron aren't absorbed as well. Vitamin C can help improve absorption.

If you decide to take iron supplements, use them as directed. Don't assume that more is better—it's not. Too much iron is toxic; in large doses it can be fatal. (In fact, iron supplements are one of the leading causes of death from accidental poisonings in the United States.)

Food Sources of Iron

Fish	Soybeans
Lean red meats	Wheat
Poultry	Black beans
Broccoli	Spinach

This table lists sources in decreasing order of bioavailability. In other words, foods at the top of the list provide the most iron in a form that the body can use; those lower on the list have less iron available in a useful form.

Some More Ways to Keep the Melatonin You've Got

Avoid Stimulants

Stimulants such as coffee or tea can affect your melatonin levels by interfering with sleep. Eliminate or reduce their use, especially in the evening. Not all sources of dietary stimulants are obvious; chocolate, for example, is a mild stimulant, and many foods and beverages contain caffeine. Check the labels.

Sources of Stimulants

- Chocolate (and foods containing chocolate)
- Coffee (and foods containing coffee)
- Cold medications containing caffeine
- Sodas containing caffeine
- Some medications, including the asthma drugs aminophylline and theophylline

Eat Regular Meals

Melatonin rhythms are strengthened by regular daily rhythms—and, as we've seen, behavioral cues are a strong zeitgeber (that is, timekeeper). So *when* you eat is just as important as *what* you eat. Regular mealtimes are important behavioral cues that keep you in sync.

Eat Light at Night

One of the effects of high melatonin levels at night is to slow your digestion process. If you eat a big meal just before you go to bed, your body won't process it as efficiently. In fact, you may find that your sluggish digestion leaves you with an uncomfortable feeling of fullness, making it hard to get a good night's rest. And your body is less likely to burn off the calories, instead converting them to fat.

Keep your eating habits coordinated with your activity levels: Eat your largest meals at times when you're most active.

A Final Suggestion

In addition to *what* you eat and *when* you eat, also give some thought to *where* you eat. The most melatonin-enhancing meal of all may well be a leisurely breakfast on the sunporch or patio. Whether you have a glass of orange or tomato juice to boost your antioxidants, milk and cereal to increase your tryptophan supply, or a low-calorie entree to stimulate your digestive system to make more melatonin, do it in the sunlight and you'll also be giving a boost to your body's natural rhythms.

Fresh fruits, fresh vegetables, and fresh air—*that's* a diet just about anyone can stick to!

CHAPTER 11

More Ways to Neutralize Free Radicals

As we've seen, the melatonin produced in our bodies is a limited resource. And as we grow older, it's important to make sure we make the most out of what we've got. One of the most effective ways of doing that is by reducing the free radicals in our bodies.

Since melatonin protects against free radicals by forming chemical reactions with them, the more free radicals that are in circulation, the more melatonin that's used up in this process. Reducing our free-radical load conserves available supplies of melatonin. In addition, by reducing free radicals, we also reduce the amount of damage our bodies' cells sustain. Even in the best of circumstances, it's not possible for melatonin to seek out and neutralize every free-radical molecule before it can cause damage. So by reducing the number of free radicals we begin with, we reduce the amount of damage we end up with.

A Free-Radical Protection Plan

There are two approaches to reducing free-radical damage: On the front end, we can limit the ways in which we're exposed to them. But since we can't eliminate all of our exposures, we should also increase our intake of antioxidants—substances that capture and neutralize free radicals within the body.

Reducing Exposures

The first line of defense is to reduce the amount of free radicals to which you're exposed. You can't, of course, eliminate all exposures; many are the consequence of daily living. For example, our bodies' cells generate free radicals when they convert food into energy. White blood cells generate them when they attack foreign intruders. Red blood cells create free radicals as a normal consequence of carrying blood throughout the body.

But you *can* reduce other sources of free radicals, particularly those free radicals that are generated by the environment. Here are some suggestions.

Stop smoking. Or at least cut back. Cigarette smoke is one of the most significant sources of free-radical exposure, especially to lung tissue. In addition, smoking depletes the body's supplies of vitamin C, a strong antioxidant. Most health plans will help you find an effective stop-smoking program, and many will even pay for it. If you're not ready to stop smoking, take extra vitamin C and other antioxidants to mitigate the damage.

Avoid pollutants. Air pollution is another major source of free radicals. Obviously you can't just hold your breath if you live in a high-pollution area, but try to limit your exposure. For example, on days when ozone levels are high, consider exercising indoors instead of running through the neighborhood. (When you exercise, you breathe in greater quantities of air.) If you run or bicycle, try to head through the park instead of along a major highway.

Air pollution can occur indoors, too. If you find that a lot of people at work experience headaches, colds, or other respiratory problems, ask your employer to have the air quality evaluated. Limit your exposure to ground and water pollution. If you work in a polluted environment, use protective clothing and air-filtering devices.

Limit chemical exposures. Pesticides (including household and garden sprays) have lots of free radicals. Don't use them if you can avoid it; when you must use them, do so responsibly. Use a respirator to protect your lungs and rubber gloves to protect your skin. Bathe thoroughly afterward, and don't eat or drink until after you've cleaned up. Be especially cautious when using pesticides in food-preparation areas. Other household chemicals—for example, chlorine bleach—also generate free radicals; use them carefully and in a well-ventilated area. Certain drugs, especially those used in the treatment of cancer, are another source of free radicals. Of course, if you've been prescribed these drugs, you need to take them. But take them only as directed.

Avoid radiation. The most likely sources of radiation are medical and dental X rays. As with medications, when X rays are needed the benefits far outweigh the risks. But try to avoid unnecessary exposures. Ultraviolet light is a less intense type of radiation, but it too can stimulate free-radical formation. Limit your exposure with sunblocks, UV-coated sunglasses, and by avoiding the use of sunlamps.

Avoid stress. As we've seen, stress interferes with melatonin's immune-enhancing properties. It also promotes free-radical formation, which may be one reason why it's often associated with heart disease and cancer.

Eat light. When you eat a heavy meal, your body temporarily shifts blood away from other organs to aid in digestion. When the blood is diverted back to other organs, it causes a rush of free radicals to be released from their tissues. (The same phenomenon happens, on a much more severe scale, when blood flow is interrupted in a heart attack or stroke. When the blood is "reperfused"—i.e., when it flows back into the tissue—it releases free radicals that can cause extensive additional damage.)

Avoid injury. Whenever your body is injured, damaged cells release free radicals. Avoid injuries by taking adequate safety precautions. If you start an exercise program, build up your conditioning gradually; aches and strained muscles are signs that you're generating free radicals. At work and at home, use proper lifting and carrying techniques so that you don't strain your back.

Design a Radical-Reducing Exercise Program

For years, you've heard that exercise is good for your health. It is. But that doesn't mean that even more exercise is necessarily better for you.

Intensive exercise may sculpt your muscles, build your endurance, and make you feel like a million bucks. But a growing number of experts are concerned that it may also cause harm. When you exercise strenuously, the activity in your muscles releases lots of free radicals, which are just as damaging as those caused by smoking, air pollution, or chemical exposure.

So while energetic exercise builds muscle mass and makes you look healthy, it may not actually make you more healthy. If you're exercising to improve your health, you're better off with a more moderate approach, one that gradually conditions your body without releasing large amounts of free radicals into your system.

"Feel the burn." "No pain, no gain." In fact, you may want to ignore those old truisms. When you "feel the burn," you're actually feeling the aftereffects of injured muscles. These and other injuries—torn knee cartilage, shin splints, tennis elbow, rotator cuff injuries—all produce huge amounts of free-radical damage.

Dr. Kenneth Cooper, the heart physician who coined the term "aerobics" and popularized the benefits of exercise, now expresses concern that overtraining may be doing more harm than good. Noting several high-profile cases of athletes who developed cancer at relatively young ages, and reports from the literature and his own patients about increased susceptibility to fatigue, infections, and injuries, he now recommends a more moderate approach to exercise.

Clearly, Dr. Cooper notes, lack of exercise is bad for your health, leaving you at greater risk for cancer, heart disease, and a range of physical ailments. But research shows that for most people, most of the benefits of exercise can be obtained without the intensive training regimens that have been so popular in recent years.

For example, a study from his Cooper Institute for Aerobics Research, published in *JAMA, the Journal of the American Medical Association,* concluded that half an hour of walking two or three times a week has almost the same health benefits as running two or more miles several times a week. The study found the biggest boost in health effects came as subjects moved from the totally sedentary group to the next most active one. But beyond that, there were only modest additional improvements in health benefits as you moved into the higher-fitness groups. Based on these and other findings, Cooper now recommends exercising at your "target heart rate" several times a week.

Calculating Your Target Heart Rate

The target heart rate is a way of gauging your exercise to be sure it's intensive enough to provide health benefits, but not so intensive that it causes damage. It's the rate at which your heart should be beating during your exercise period. Here's how to calculate it:

Begin by figuring out your *maximum heart rate.* That's the number of heartbeats per minute when you're exercising as hard as possible. But you don't have to go out and run wind sprints to find out what it is; you can use a simple formula instead. Simply subtract your age from 220. So if you're 50 years old, your maximum heart rate is 170. Although the actual rate varies a bit from person to person, this formula gives a good approximation.

Next, multiply by 0.7 to get an approximation of your target heart rate. In our example, 170 × 0.7 = 119. So your target heart rate will be approximately 119 beats per minute.

As you exercise, your actual heart rate should approximate your target heart rate, within about ten or fifteen beats in either direction. While you're exercising, stop briefly and take your pulse. It's easiest to take your pulse for fifteen seconds and then multiply it by four rather than counting an entire minute.

The main thing to keep in mind is not to overdo it. If your heart rate during exercise is consistently more than ten or fifteen beats higher than your target rate, you may be generating excessive free radicals.

Type of exercise. Any kind of exercise that brings your heart rate up to the target rate will be effective. *Low-impact* exercise, however, will help prevent the kinds of sports-related injuries that contribute to free radical damage. Also, keep in mind that the exercise should be continuous. Sports such as baseball and football where you have bursts of activity mixed in with periods of inactivity aren't as beneficial as sustained movement. If you play tennis, singles will give you a more consistent workout than doubles, while calisthenics or jogging in place can keep your heart rate up in between points. Swimming is good exercise, but it's hard to keep your heart rate high enough; as you swim more, you become more efficient at it and your body doesn't have to work as hard. Fast walking is one of the best forms of exercise, and even household chores like raking, vigorous yard work, and chopping firewood can be effective.

Your target heart rate won't change as you become more fit, but you'll have to work harder to hit it. As your body adapts to your exercise regimen, the heart becomes more efficient, so you'll have to push a little harder to bring the heart rate up. As your program progresses, use the target heart rate as your guide to increasing your activity. When you become better conditioned, you'll be able to exercise more vigorously without raising your heart rate too high.

The amount of exercise you do is really a personal choice. At a minimum, you should exercise at least twenty minutes three or four times a week, and this exercise should be vigorous enough to keep your heart rate at or near the target heart rate for the entire twenty minutes. Some studies suggest that you can get some health benefits from even less exercise, but this is generally accepted as the minimum you need for cardiovascular conditioning. Of course you can exercise more—and most people eventually do increase their exercise as they continue with their programs. But the key to preventing free-radical damage is to build up *gradually* so that you don't damage your body and generate free radicals.

CHAPTER 12

Zapping EMFs

We saw in chapter 7 how electromagnetic fields (EMFs) have a powerful effect on the pineal gland, sometimes shutting down production of melatonin almost completely. In this chapter, we'll offer some concrete strategies for reducing your exposure to EMFs. These strategies are built around the concept of prudent exposure introduced earlier. EMF exposure is an inevitable consequence of modern living, and it's impossible to reduce your exposure to zero. But you can take steps to reduce the amount of EMFs you absorb.

Current evidence also suggests that all EMFs are not created equal, and it makes sense to zero in on the most harmful types first. If the theories are correct, most if not all of EMFs' health effects result from their action on the pineal gland. And if that's so, then it follows that we should look first to limiting *chronic* exposure to EMFs. As we've seen, the pineal gland's melatonin production bounces back quickly after EMF exposure stops. But

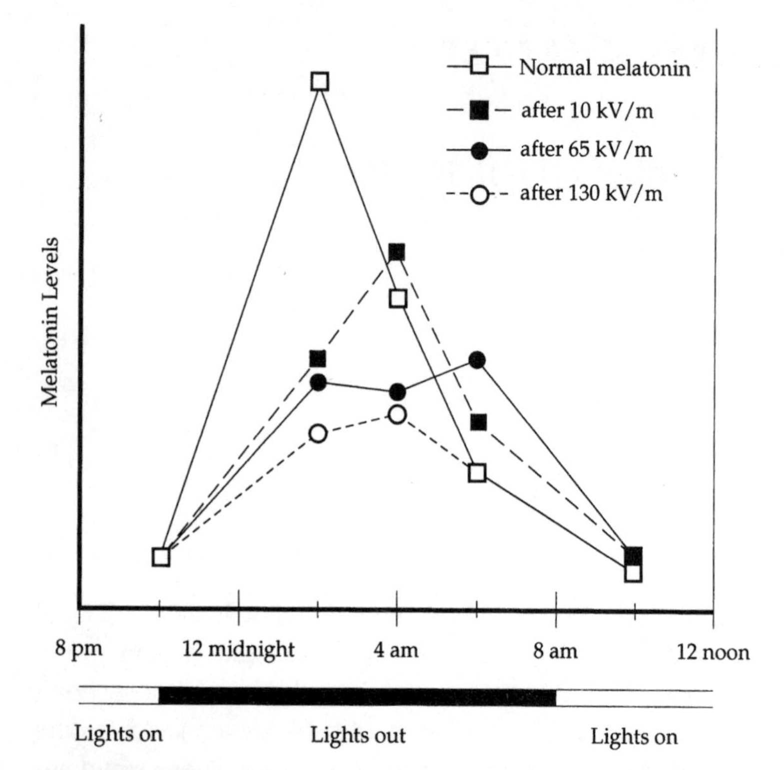

How EMFs Affect Melatonin Levels. *Electric fields exert powerful influences on pineal gland activity, depressing melatonin levels in immature rats. Scientists believe that humans are affected by EMFs in much the same way.*

chronic exposure—even at lower levels of intensity—never gives the pineal a chance to recover. In this respect, EMFs act just like light. Intermittent brief exposures to bright light appear to have less effect on pineal function than ongoing exposure to lower light levels.

The following table provides a graphic example of the differences in EMF emissions. Compiled by the U.S. Environmental Protection Agency, this list shows typical levels of EMFs from a variety of household sources and appliances. But as you can see, the realities of EMF exposure get pretty complicated. Some appliances, for example, emit high levels of EMFs, but we may be exposed to them for only a few minutes a day. Also, the exposure levels change dramatically depending on where you are in relation to the source. EMFs will be high within a few inches of a small but intense source, but at a distance of just one or two feet they're negligible. By contrast, larger sources—such as house wiring—may generate lower-strength fields but over larger areas.

So with these thoughts in mind, our approach to reducing EMF exposure is based on the following principles of "prudent exposure":

- Begin by making changes that will produce the most benefit with the least disruption—in other words, those changes that offer the "biggest bang for the buck." For example, consider changing your appliances before you change your address.
- Look first to reduce or eliminate chronic sources of exposure rather than intermittent sources. For example, take a closer look at the EMFs from your electric blanket than your electric hair dryer.
- Finally, use as a working rule of thumb the general proposition that any reduction in exposure will be beneficial, and that the greater the reduction, the greater the benefit.

EMFs Emitted by Household Appliances and Devices

	Milligauss at:			
Device	*½ ft*	*1 ft*	*2 ft*	*4 ft*
Air cleaners	180	35	5	1
Baby monitors	6	1	0	0
Battery charger	30	3	0	0
Blanket, electric (older models)	200+	NA	NA	NA
Blanket, electric (reduced-emission models)	20	NA	NA	NA
Blender	70	10	2	0
Can opener, electric	600	150	20	2
Ceiling fan	NA	3	0	0
Clock, dial	NA	15	2	0
Clock, digital	NA	15	2	0
Clothes dryer	3	2	0	0
Coffeemaker, electric	7	0	0	0
Computer monitor, color	14	5	2	0
Copier	90	20	7	1
Crock Pot	6	1	0	0
Dishwasher	20	10	4	0
Drill, electric	150	30	4	0
Fax machine	6	0	0	0
Fluorescent lights	40	6	2	0
Food processors	30	6	2	0
Garbage disposal	80	10	2	0
Heater, portable	100	20	4	0
electric	200	70	20	2

Iron	8	1	0	0
Mixer, electric	100	10	1	0
Oven, electric	9	4	0	0
Oven, microwave	200	40	10	2
Range, electric	30	8	2	0
Saw, power	200	40	5	0
Stereo	1	0	0	0
Television, B&W	NA	3	0	0
Television, color	NA	7	2	0
Toaster	10	3	0	0
Typewriter	100	1	0	0
Vacuum cleaner	300	60	10	1
Washing machine	20	7	1	0

"0" indicates that field strengths could not be distinguished from background levels. "NA" indicates that no measurements were available.

Source: Environmental Protection Agency; reported in Pinsky, Mark A. *The EMF book.* New York: Warner Books, 1995, p. 44.

The last point seems obvious, but we really must take it on faith. In fact, there are probably upper and lower thresholds of exposure. In other words, if a given level of EMF exposure shuts down melatonin production altogether, it wouldn't matter how much more exposure you get on top of that. Similarly, there may be a minimum threshold—as with light—below which EMF exposure doesn't appreciably impair melatonin production. But we are a long way from knowing if such lower and upper thresholds exist, much less what they might be. (As we saw in chapter 7, the current recommended levels of "safety" aren't based on measurable health effects; rather, they simply reflect typical levels that most people are exposed to.) There are also great difficulties in getting accurate measurements of EMF exposure for any given individual. Two people living in the same house or working at the same office may receive far different exposures. Given all these uncertainties, the most prudent and reasonable course is to eliminate or reduce EMFs wherever it is practical to do so.

Following these three general principles, you should concentrate first on identifying and reducing chronic exposures and then on high-intensity but short-duration exposures.

Measuring EMFs

Further compounding the problem is the difficulty of measuring or estimating your exposure to EMFs. You can buy a *gauss meter* to measure levels in your home. A low-end meter costs about $100, but it may give inaccurate readings. More sophisticated models are accurate, but require special training. (To get a list of meters for consumers, send one dollar to Microwave News, P.O. Box 1799, Grand Central Station, New York, NY 10163.) Your electric utility may have a program to measure EMF exposures in your home, or it may be able to recommend a consultant who can do the testing; call and ask.

These measurements provide only a rough guide to your exposure. For example, EMF levels from utility wires and household wiring vary widely by time of day and season as current loads change. And they vary by location: They may be strong on one side of a room and minimal just six feet away. Also, power lines aren't the only source; exposure from appliances may constitute a greater portion of our total EMF exposure. More important than absolute numbers are the ranges. If you find high values, repeat the tests at different times of the day. Look especially for hot spots (e.g, near distribution lines).

The *level* of exposure is only part of the story, however. You also need to look at how *chronic* the exposure is. As animal studies show, melatonin levels bounce back quickly after EMF exposure, so the issue may not be how high the levels get so much as how long they last. While it's true that appliances such as portable hair dryers and cellular phones may expose users to much higher levels of EMFs than power lines, these exposures usually don't last very long. Exposure from electric wiring (both outside and inside the home), by contrast, tends to be ongoing and continuous.

Reducing Chronic Exposure

Power Line Exposures

Much of the initial focus on EMFs involved exposures from power lines and associated equipment. Although we now know that other exposures may be just as or more significant, power line exposure continues to be a concern, especially because individual consumers can do little to alter their exposure short of moving.

For many people, the most visible source of power line exposures are *distribution lines*—typically, three wires that are strung on a telephone pole. These lines generally carry current for households and light commercial facilities. In general, older lines may be more likely to generate EMFs because they're more likely to be unbalanced. In its simplest terms, balanced loads mean that incoming and outgoing currents are closely matched. In many cases, older lines tend to become unbalanced.

Besides balance, another issue in EMFs associated with electrical lines is the concept of *phasing*. Power lines are more likely to emit fewer EMFs when the lines carrying inbound current run *close to* and *parallel* to those carrying outbound currents. As a result, the fields they generate tend to cancel one another out.

One reason EMFs may be higher when wiring is older is because they get out of phase: For example, as wires sag or utility poles shift, the wires may no longer run parallel. A tree branch pressing on one of the lines may also make them nonparallel.

Buried lines are likely to emit fewer EMFs—not because they're underground (which doesn't do much to block the fields) but because they're likely to be newer and are less likely to shift position. Newer lines may also use twisted cables, in which the three lines are twisted around one another into a single larger cable. This configuration usually reduces EMFs dramatically, because the wires automatically stay in phase. If you have overhead wires, look to see if there's one fat wire in place of three skinny ones.

Other power line sources of EMFs include *high-tension lines, transformers,* and *substations*. Take a look around your neighborhood to see if there are any nearby. High-tension lines—also known as transmission lines—carry high-voltage currents from the power plant to local substations, usually along giant gantries that are easily visible. Depending on their design and construction, EMFs from such lines vary widely, and there's no universally accepted standard for "safe" distance between them and nearby houses. In some cases, EMF levels above 2 milligauss have been measured as far as 300 feet from such lines. If your home is farther away than that, chances are your exposure is minimal; if it's closer, you may wish to have field levels inside your house measured.

Substations and transformers both "step down" electricity from a higher to a lower voltage before it goes to individual houses. Substations are usually recognizable by the chain-link fences and high-voltage warning signs that surround them, as well as by the Dumpster-like power transformers inside and the wires converging on the site. (In cities and some upscale neighborhoods, they may be harder to spot; they may be located, for example, in the basement of an office tower or in a building designed to look like houses in the area.) Smaller distribution transformers often look like metal trash cans hung on a utility pole; in communities where wires are buried underground, however, the distribution transformers will be on the ground. Typically, they're housed in green metal cabinets a little bigger than a washing machine, mounted on a concrete pad, with a yellow high-voltage warning label on the outside.

What you can do about power line exposures. If you suspect high EMF emissions, your local utility may be willing to replace older wiring systems, especially the lines going directly to your house. In many cases, these wires are due for replacement periodically anyway due to normal wear and tear.

The farther up the distribution network you go, the harder it will be to get action from your utility. The utility may be willing to measure EMFs in your home if, for example, there's a distribution transformer directly outside your child's bedroom window. It may, under some circumstances, even be willing to relocate the equipment if levels are high. But when you request that substations be moved or high-tension wires rerouted, that's another proposition altogether. Utilities have millions of dollars invested in such installations and equipment. Politically the stakes are high, too, and many utilities will be unwilling to concede to such demands because of fears that they will be inviting lawsuits, will incur higher equipment costs, or will be unable to build new transmission and distribution networks and facilities at all. In such cases, action is most likely to be effective if it is requested through consumer groups or the state Public Utilities Commission. If you believe you're at risk from utility-generated EMFs, by all means add your voice to the debate. But be prepared for a long and difficult struggle. And in the meantime, take additional measures to reduce your exposure from other sources at home and work.

Household Wiring Exposures

The technical issues surrounding EMFs from household wiring aren't much different than those for power line exposures. If you own your home, of course, you have one key advantage in addressing them: You own the equipment and you can change it. (That's also a disadvantage, of course, since these changes may be expensive.)

Compared with power lines, household wiring is less likely to be adequately balanced, especially in older homes. Electricians try to balance circuits when they install the wiring, but results may vary widely from house to house. Later wiring changes—the addition of a new outlet or lamp, for example—may unbalance the original scheme, especially if the work wasn't done by a licensed electrician. Installation shortcuts may also increase EMFs—for example, if the household wiring is grounded to the plumbing rather than directly to the earth, or if current "leaks" from poorly insulated wiring. An electrician can test your wiring to find out if it's balanced.

Household and office wiring is often out of phase as well, even if it was designed to be in phase originally. Rewiring a room to allow the lights to be turned off and on from more than one wall switch, for example, can throw the room's wiring out of phase and dramatically increase the EMFs given off.

In newer homes, wiring is kept in phase because both the incoming and return wiring are wrapped in a single cable, where they cancel each other out. An electrician may be able to fix out-of-phase wiring if it's caused by improper connections or grounding.

To determine your risk from unbalanced and out-of-phase wiring, you can measure EMFs directly inside your home. Alternatively, a little detective work will allow you to uncover and correct the most likely sources of high EMF exposure from your household wiring. Here are the most common culprits:

Improperly installed electrical wiring. As we noted, a licensed electrician can usually spot such problems as improper groundings and out-of-code wiring arrangements. He or she may also have the equipment to measure EMFs directly. You can spot some of these problems yourself. Look to see where the grounding cable from your circuit box is located. It should be attached to a metal spike that goes directly into the ground. If it's attached to a pipe, your entire house's plumbing system may be generating EMFs.

If circuits or switches have been added, they may have put your wiring out of phase. Likely culprits are two-way wiring schemes that allow a light to be turned on or off from more than one switch, and nonprofessional installations—for example, the installation of a new outlet or switch.

Dimmer switches. Dimmer switches are also a big source of EMFs. They convert some of the line voltage into EMFs; the dimmer the light, the stronger the field that's thrown off. It's fairly easy to replace them if you wish to (just make sure you turn off power at the circuit box first!), but you can also just keep the lights turned up to reduce EMF exposure.

Exposure from Electrical Appliances

Many electrical appliances generate high levels of EMFs when they're in use. A few of them may create chronic exposures.

Electric blankets. Because of the way they're constructed and used, electric blankets have raised special concerns as a potential source of EMFs. We're in contact with them for lengthy periods of time; they're used in close proximity to the body, and they "blanket" us within a steady field of EMFs. In addition to its length, this nighttime exposure may be especially risky, since that's when your body manufactures melatonin.

In one study, researchers found that the *type* of electric blanket makes a big difference in the effects on the pineal gland. They compared conventional electric blankets with a type known as "continuous polymer wire"

(CPW) designs. The latter are designed in a way that makes them temperature sensitive, applying more heat to cooler parts of the body. These blankets emit up to 50 percent more EMFs than conventional blankets; further, they switch on and off about twice as often per hour. Conventional electric blankets showed no measurable effects on the function of the pineal gland. CPW blankets, on the other hand, significantly reduced the production of melatonin. It may be this switching—and the electric currents it creates in body tissues—that depresses melatonin production.

One suggestion to reduce EMF exposure is to use the blanket to preheat your bed, but shut it off when you get in (EMFs don't "linger"; once the source is shut off, they disappear). Another idea: Replace your electric blanket with a down comforter—you'll also save on your electric bill.

Alarm clocks. The electric motors in dial-face clocks can emit EMFs. Like electric blankets, alarm clocks are a special concern because we tend to sleep next to them, and for a long period of time. If you sleep within one foot of such a clock, move it to another location or replace it with a digital clock. Digital clocks usually have much lower levels of EMFs.

Waterbed heaters. Electric heaters for waterbeds pose the same issues as electric blankets, though they usually create weaker fields. Again, the biggest concern is that you're exposed to these fields continuously as you sleep.

Lighting. Fluorescent lights are a source of EMFs, though their intensity drops off with distance. Halogen lights usually have a small transformer in the base, which creates fewer EMFs. Incandescent lights create even fewer EMFs; use them when you have a choice.

Computer monitors. Computer monitors are widely linked in the public mind with EMFs, but as the table above shows, they generally emit about the same or lower field strengths as electric typewriters. Even so, it's wise to limit your exposure. Distance is your best defense; field strengths drop off rapidly. In most cases, the strongest fields aren't directly in front of the terminal; they're to the back and right (because of the location of the transformer). If you work in a crowded office, look around; you may be getting more exposure from a coworker's terminal than your own—even if it's on the other side of a cubicle or office wall. If you can, move your chair or ask your neighbor to move his or her monitor. If you use a laptop computer, you're in luck: Flat-panel displays don't emit significant levels of EMFs.

Copiers and printers. Photocopiers and laser printers both emit strong EMF pulses when they print a page. As with many EMFs, the effects diminish rapidly within a few feet, so try to keep your distance. For example, if you're making numerous copies, stand away from the machine while it's running. If your laser printer is next to your chair, move it a few feet away.

Commuting. Most electric trains and subways run on direct-current (DC) power, which produces much lower levels of EMFs than alternating current. Even so, commuter trains often contain onboard electrical devices that can create "very intense" fields, according to the U.S. Environmental Protection Agency. If you have a gauss meter, you can use it to gauge exposure in different parts of the railcar on your next trip, and plan your future commutes accordingly. If, for example, seats near the doors consistently show high EMF levels, try to sit in other seats during your daily commute.

Hidden sources. In addition to these sources of EMFs, you should also consider whether you might be exposed to any hidden sources. Remember, EMFs travel easily through walls, ceilings, and floors, so think about sources you can't see directly. For example, large commercial air-conditioning equipment is a source of EMFs. If you work directly below such equipment (which is often located on the roof of multistory office buildings), you may be exposed to the field. Distribution transformers are often located in basements or near electrical equipment such as mainframe computers. If you work in a hospital, find out if nuclear magnetic resonance (MRI) equipment is close by; it also generates strong EMFs. If your office is near the company kitchen, you may be getting EMFs from the vending machines (which often contain fluorescent lights). At home or at work, find out where the circuit box is located; it too emits strong but limited fields. If, for example, the circuit box is in a living room closet next to your favorite chair, consider rearranging your furniture.

Intermittent Exposures

To reduce your exposure from appliances and other intermittent sources of EMFs, you can usually make minor changes in how you work. For example, standing away from a photocopier while it's running will reduce your EMF exposure.

As the table on pages 151–152 shows, there are numerous intermittent sources of EMFs in the workplace—some under your control, some not. For example, fluorescent lights in an office *beneath* yours will create EMFs that travel through the floor; however, there's usually not much you can do about such exposures. The speakers in telephone earpieces emit EMFs, but unless you prefer face-to-face meetings there's probably not much you can do about them.

Appliances such as electric pencil sharpeners and postage meters have high-speed motors that create strong fields; however, the intensity of such fields usually drops off rapidly with distance, and moving a foot or two away from them usually minimizes your exposure. Review the table on pages 151–152 to identify these and other potential sources and look for ways to eliminate or reduce your exposure. (For example, move the department's electric pencil sharpener off your desk, or replace it with a manual model.)

Look especially for ways to reduce your exposure to pulses—short bursts of energy that cause rapid fluctuations in EMFs. Turning equipment on or off causes a pulse, so you may wish to keep your computer on even if you step away from your desk for a while.

This chapter has given you a brief overview of the sources and hazards of EMFs. A good source of further information is *The EMF book* by Mark A. Pinsky, or you can obtain more information at your local library. The important thing to remember is that EMFs are one of the most significant inhibitors of melatonin production; the less exposure to EMFs you receive, the more you'll be protecting your body's natural melatonin supply.

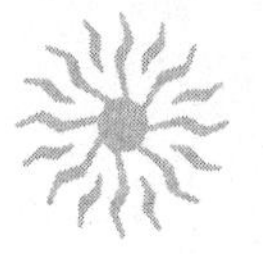

CHAPTER 13

Checking Out the Medicine Chest

The pineal gland sits at the crossroads of the nervous, immune, reproductive, and endocrine systems, helping to orchestrate the vast and complex processes of daily living. So it's not surprising that when we use medications to influence some aspect of that system, we often affect the pineal in the process. Medications can interact with melatonin directly and indirectly. And not all of these interactions are for the worse. For example, melatonin may enhance the effectiveness of certain drugs, permitting lower doses.

Because so much of the research on melatonin is new, we are just beginning to understand how medications affect and are affected by the pineal gland and melatonin. But to the extent that melatonin helps keep our natural rhythms strong, fights infection, cancer, and free-radical damage, and slows the aging process, chances are we'll be hearing lots more about its role in treating disease in the months and years to come.

Alpha- and Beta-Blockers

The drugs with the greatest impact on melatonin production are those known as alpha-adrenergic and beta-adrenergic blocking agents (alpha- and beta-blockers for short). These drugs are used to treat a variety of conditions; their most common use is for treatment of heart disease and hypertension.

Alpha- and beta-blockers can virtually shut down melatonin production by short-circuiting the link between the eyes and the pineal gland. In fact, many of the side effects of such drugs, which may include sleep problems, mood disorders, disorientation, memory loss, fatigue, and visual disturbances, are similar to the effects of altered melatonin levels.

These drugs work by affecting "receptor" sites on cells. You will recall from chapter 2 that cells communicate with one another by chemical messengers known as neurotransmitters. These messengers fit into specific receptors, ensuring that the right message gets to the right cell. You can think of the neurotransmitters as "keys" and the receptors as "locks." When the right key fits the lock, the engine starts.

Alpha- and beta-receptors are sort of general-purpose locks, found throughout the body to control various systems. In the heart, for example, beta-receptors help regulate the rate and strength of the heartbeat. Alpha-receptors help blood vessels know when to expand and contract to control blood pressure.

Sometimes, though, these receptors are too sensitive to neurotransmitters, or the neurotransmitters are too active. That's when blocking drugs are used. The drugs are designed to fit into the receptors, preventing the neurotransmitters from turning them on.

To understand how these drugs affect the pineal gland, we must look back to how the pineal "sees" light and dark. You will recall that a message travels from the retinas along a complex pathway of nerves and arrives at the pineal gland.

The pinealocytes—the groups of cells within the pineal gland that actually manufacture melatonin—contain both alpha- and beta-receptors. These receptors are the final link between melatonin manufacture and the light-dark signals traveling from the eyes. When these signals reach the "end of the line" on the neural network, they release neurotransmitters that act on these receptors to start up and shut down the production of melatonin.

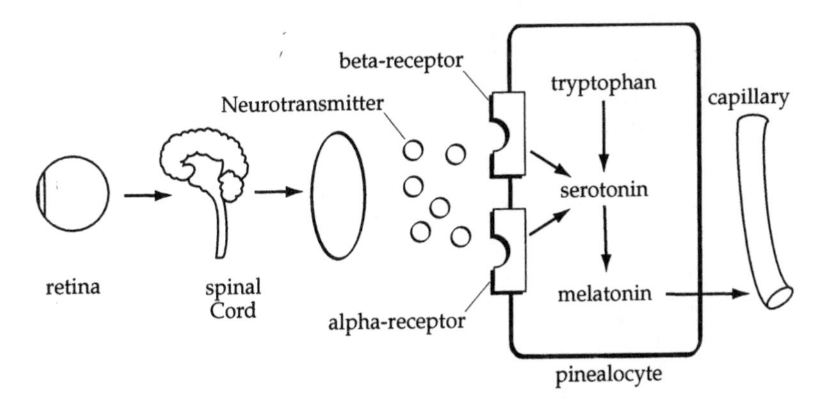

Alpha- and Beta-Receptors and Melatonin Production. *Signals traveling from the eyes stimulate the release of neurotransmitters, which act on cell receptors in the pineal gland to turn melatonin production on and off.*

When we take alpha- and beta-blockers to treat heart disease, hypertension, or other conditions, they also make their way to the pineal gland, where they block the light-dark signal from getting through. The result: Melatonin production is significantly or completely impaired.

The following table lists some common brand names for alpha- and beta-blockers. (The names of generic versions often end in "-amine" for alpha-blockers or "-olol" for beta-blockers.) If you're taking drugs for hypertension and/or heart conditions, ask your doctor if any of them are alpha- or beta-blockers.

Common Alpha- and Beta-Blockers

ALPHA-BLOCKERS	
Hytrin	Priscoline
Minipress	
BETA-BLOCKERS	
Blocadren	Normozide
Cartrol	Sectral
Corgard	Tenoretic
Inderal	Tenormin
Inderide	Timolide
Levatol	Visken
Lopressor	
ALPHA- AND BETA-BLOCKERS	
Normodyne	Trandate

If you're taking these drugs, don't discontinue them or change the dosage on your own. In heart patients especially, these drugs are essential and you could severely damage your heart—or even die—if you change the dosage without your doctor's supervision. Always consult with your doctor before making any changes in your dosage.

However, in light of the facts that these drugs are known to impair melatonin production, and that melatonin, in turn, helps protect the heart, discuss with your doctor the possibility of adding melatonin supplements to your treatment regimen (see chapter 15).

Further, ask your doctor if your dosage can be adjusted so that nighttime levels are reduced. Since melatonin is produced mostly at night, reducing nighttime levels of these drugs can help keep your rhythms on track. As an added benefit, it may help reduce or eliminate many of the side effects of these drugs.

Melatonin-enhancing lifestyle changes are important too, but since alpha- and beta-blockers shut down the light-dark signal at the pineal gland, lifestyle changes can't make up for their melatonin-inhibiting effects.

NSAIDs

A class of drugs known as "nonsteroidal anti-inflammatory drugs" (NSAIDs) may also inhibit your melatonin production. Two of them in particular—good old aspirin and indomethacin—have been shown to inhibit melatonin levels. Since most NSAIDs are chemically related, other drugs in this category, such as ibuprofen and naproxen, may inhibit melatonin as well.

If you're taking NSAIDs for pain (as opposed to inflammation), consider switching to acetaminophen (for example, Tylenol). It has a different chemical structure, and there's some evidence that it doesn't affect melatonin production. While acetaminophen is as good as aspirin for pain, it doesn't relieve inflammation, so it's usually not a good choice for arthritis and other inflammatory diseases. If you must take NSAIDs, consider taking melatonin supplements as well (see chapter 15).

Other Drugs Affecting Melatonin

Drugs Affecting Sleep and Activity Cycles

Stimulants. Many other drugs can affect melatonin production by disrupting your body's natural rhythms. For example, stimulants ranging from caffeine to amphetamines interfere with sleep patterns and throw your melatonin production off. In one study, caffeine was also found to have direct effects on the pineal gland: When it was combined with low doses of the beta-blocker propranolol, it blocked melatonin production in rats. Neither substance given separately reduced melatonin levels. So if you are taking beta-blockers, it's especially important to forgo coffee and other caffeine-containing foods and beverages.

Sedatives. Sedating drugs, such as barbiturates, sleeping pills, and alcohol, are also disruptive. Even though they may help you *fall* asleep, they don't let you get the right *kind* of sleep. For example, many of them interfere with the rapid-eye-movement (REM) phase of sleep—the dreaming stage that's essential for complete rest. That's one reason sedatives often leave you feeling less rested the following day.

Benzodiazepines. Such drugs as Valium, Librium, and Xanax appear to have a close chemical relationship with melatonin that may help explain their antianxiety and "hypnotic" properties. These drugs act on many of the same systems as melatonin itself, and may mimic its sleep-promoting properties.

In fact, animal tests show that melatonin can enhance the effects of benzodiazepines. In one experiment, mice were treated with a combination of melatonin and low doses of diazepam (Valium). On its own, the diazepam dose wasn't large enough to affect behavior; when it was combined with melatonin, however, it produced significant clinical effects.

If you're taking benzodiazepines, these findings have some practical implications. First, by increasing your melatonin supply—through lifestyle changes, supplements, or both—you can reduce your use of these powerful and habit-forming drugs. For example, by bringing your activity cycles in line with natural cycles of light and dark, you will have less need for medications to help you fall asleep. (Consult with your physician, of course, before making any changes.) And as your melatonin rhythms grow stronger, you will likely find that reduced doses will still give you the same therapeutic effect. In fact, if you *don't* reduce your dosage, you'll probably find yourself *overmedicated* as rising levels of melatonin enhance the drug's effects.

In addition, if you now rely on benzodiazepines to help you sleep—as millions of Americans do—melatonin may help you kick the habit. By interfering with natural sleep cycles, these drugs disrupt normal melatonin rhythms. These disruptions, in turn, may reinforce the feeling that you need the medication. In other words, the reason you need to take these drugs to get to sleep is because these drugs are *interfering* with your sleep.

Improving your natural melatonin rhythms may help reverse this vicious circle: As your sleep cycles become more normal, you won't be as dependent on benzodiazepines or other artificial sleeping aids. Then, as you take less of these drugs, your sleep cycles will improve even more, further reducing your need for drugs . . . and so on. If you take Valium or other benzodiazepines regularly (for example, more than once a week), ask your doctor to help you design a melatonin-centered regimen for reducing their role in your life.

Marijuana. There have been no reported studies of research on the effects of marijuana on melatonin production, but there is some evidence that it acts along the same pathways as benzodiazepines. To the extent that it does, the same issues apply. In addition, marijuana and related drugs (primarily hashish) are known to affect the immune and reproductive systems and cause sleep disturbances—all implying a link between marijuana use and impaired melatonin levels.

Other psychoactive drugs. Virtually all drugs used to treat psychiatric and mood disorders have some effect on circadian rhythms. Some, such as antidepressants, may help bring abnormal rhythms into line; others may throw them out of whack. If you or a loved one are taking these medications, ask your doctor about their effects on sleep and activity cycles, and explore whether melatonin might prove helpful.

Drugs Impairing the Immune System

A wide variety of drugs interfere with the immune system, placing added burdens on melatonin's immune-enhancing properties. In particular, corticosteroids and certain anticancer medications tend to blunt your immune system. While these drugs don't seem to affect melatonin levels directly, they do make it harder for melatonin to do its job. If you're taking these medications, ask your doctor whether they affect your immune system. If they do, discuss with your physician taking melatonin supplements to give your system an added boost (see chapter 15).

Drugs Affecting Endocrine/Hormone Activity

Similarly, any drugs that work by affecting the endocrine system or hormonal levels may disrupt your body's natural rhythms and communication systems. Such medications include birth control pills, estrogen replacement therapy, ACE inhibitors (a type of heart medication), high blood pressure medications, thyroid pills, and antidiabetic medications—to name just a few.

There's no hard-and-fast rule about how such drugs might affect your body's systems and rhythms, and few studies have been conducted on how such drugs affect the pineal gland. But if a drug's side effects include sleep disturbances, impaired immunity, mood shifts, or fatigue, there's a good possibility that it may be affecting melatonin levels.

Melatonin's Effects on Medication

For all these drugs, the route to the pineal may be a two-way street. As we've seen, melatonin has been shown to enhance the effects of some drugs; it may affect others in similar ways.

If you're on medication and you begin taking melatonin supplements, let your doctor know. He or she may wish to monitor the effects of your medication and consider adjusting the dosages up or down if there are any changes.

Similarly, as you implement strategies to boost your natural melatonin levels, you may improve the effectiveness of medications you're currently taking. Ask your doctor to consider reducing dosages on a trial basis, especially for medications that melatonin is known to affect (for example, benzodiazepines, antidepressants, estrogen replacement therapy).

Serotonin Uptake Inhibitors

If you're taking a class of drugs known as *serotonin uptake inhibitors*—which is used to treat anxiety and certain other psychological disorders—you should *not* take melatonin supplements. Because of the relationship between serotonin and melatonin, supplements can interfere with the action of these drugs. The best known of these drugs is Prozac; however, new medications and formulations may come onto the market, so if you're taking any medications for obsessive-compulsive disorders, mood disorders, or anxiety disorders, check with your physician before taking melatonin supplements.

Recommendations

To help ensure that your medications and melatonin are working together:

1. Check your medications. If you're not sure, ask your doctor or pharmacist whether you're taking NSAIDs, beta-blockers, serotonin uptake inhibitors, alpha-adrenergic blockers, benzodiazepines, or other drugs that affect melatonin. If you are taking such drugs, ask your doctor about switching medications, adjusting dosages, or adjusting your dosage schedule (for example, taking the medications in the morning versus the evening).

2. Watch for melatonin-related side effects. If you have trouble sleeping, if you seem to catch a lot of colds, or if you're tired or depressed, melatonin-depleting medications may be the reason. Notify your doctor about the problems you're experiencing, and ask whether he or she can recommend a different medication or dosage.

3. Look at nonprescription and "hidden" drugs. Examine your use of nonprescription drugs, such as aspirin, cold medications (many of which contain caffeine), and foods and beverages such as coffee and chocolate.

Always check with your doctor before making any changes in the dosage of your medication. But by strengthening your body's melatonin supply, you may be able to offset some of the negative effects of medications, or in some cases reduce your need for them.

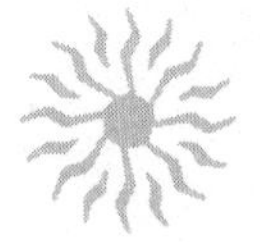

CHAPTER 14

Melatonin and Jet Lag

As the role of melatonin in regulating circadian rhythms began to become clear, it occurred to physicians that this ancient substance might be useful for treating a most modern disorder: jet lag.

When the term was first coined, jet lag was seen as an unfortunate but unavoidable consequence of an exotic lifestyle, afflicting the small world of privileged "jet-setters" who could afford to fly from coast to coast. Confined to this exclusive circle, it wasn't a problem that attracted a lot of attention in medical circles, and among world travelers it was often seen as almost a badge of honor, a minor annoyance that was part of the price for a trendy lifestyle.

Today, as business goes global and airplanes fill the skies, jet lag seems a lot less exotic. As the ranks have grown with people who must regularly travel across the country and around the world, we're learning that jet lag is more than a medical curiosity. It's an occupational hazard for thousands of workers. It reduces productivity and is a likely contributor to many cases of travel-related accidents and misjudgments. It is suspected, for example, that many cases of pilot error may be attributable, at least in part, to jet lag. And we now know that jet lag, and its more proletarian cousin shift work, can also contribute to long-term health problems such as impaired immunity, high blood pressure, and heart disease.

From a physiological perspective, there's little difference between jet lag and shift work. Both involve a rapid change in the body's rhythms, shifting them forward or backward by several hours or more. Until the body adapts, we face the problem of trying to function while the body "thinks" it should be sleeping, and trying to sleep when the body's prepared for activity. These effects can't be overcome simply by getting more sleep; they require the body to reset its internal clock—a process that can take up to twelve days, depending on the extent of the shift.

For frequent travelers or those working erratic shifts, the problem is compounded further. Just as the body begins to accommodate itself to the new routine—a process that takes several days or longer—one's schedule either changes back to the original schedule or to another one altogether. For some people, the changes occur so frequently that the body never really gets a chance to regain its equilibrium.

Over the years, doctors have tried many strategies for treating jet lag and shift work. Until recently, those efforts have met with limited success. Often the best advice that could be given was to give your body time to adjust—a good recommendation, but not always practical.

Then researchers who had been working with

melatonin began to consider its potential as a treatment for the problems of jet lag and shift work. Beyond their practical implications, these investigations were important because they opened the door to a whole new approach to understanding disease. Known as *chronobiology,* this approach looks at a variety of health problems and considers how disruptions of normal rhythms might play a role.

One of the most significant milestones in this new field was the publication of studies by Alfred Lewy, M.D., Ph.D., a psychiatrist at the University of Oregon, showing that it was possible to mitigate the effects of jet lag with both melatonin supplements and light therapy. The findings, which also apply to shift work, offer practical benefits, but they're also significant because they demonstrate both the importance of melatonin rhythms to our health and feelings of well-being, as well as the ability to modify those rhythms.

Shifting with the Sun

One of the best things about Lewy's findings is that they offer some simple strategies for adjusting to new circadian rhythms. While supplements can help with this transition, Dr. Lewy doesn't recommend melatonin supplements, except under the supervision of a physician, because of the risk that incorrectly-timed doses can throw the internal clock off even further. He prefers an even simpler treatment: daylight. "Light works very well," he says, and it carries less risk of inadvertently setting your internal clock to the wrong cycle.

In a study published in the *Psychopharmacology Bulletin,* Dr. Lewy and coauthor Serge Daan, Ph.D., offer the following recommendations for treating jet lag and shift changes:

1. Use daylight, if at all possible, to reset your rhythms. To affect pineal production of melatonin, you need light with intensities of more than 1,000 to 2,000 lux (a lux is a unit of light exposure; it depends both on the brightness of the source and how close you are to it). Indoor lighting is generally in the range of 200 to 500 lux. Outdoors, a cloudy day has a brightness of 10,000 lux; bright daylight is approximately 100,000 lux.

2. When flying from *east* to *west* (e.g., New York to Los Angeles), *avoid morning* exposure to daylight in the days after you arrive. *Seek out* daylight exposure in the afternoon.

This approach works as long as you're not traveling across more than 6 time zones. If you're traveling 6 to 12 time zones westward, however, get your light exposure *early* in the afternoon. Avoid late afternoon exposures; by this time, your body will be fooled into thinking it's being exposed to morning light, not evening light.

If you're traveling more than 12 zones westward (i.e., more than halfway around the world), it's equivalent to traveling eastward from a circadian point of view. For example, traveling 16 zones to the west will land you in the same time zone as if you'd traveled 8 time zones eastward. So consider any westward travel across more than 12 time zones as eastward travel (see next step).

3. For eastward travel (Los Angeles to New York), simply reverse the preceding instructions. *Avoid* expo-

sure in the evening and *seek* it in the morning. If you're traveling from 6 to 12 time zones eastward, avoid *early* morning light; get exposure to late morning light instead.

Go to bed at the regular bedtime *at your destination*—or as close to that time as is practical. Likewise, try to get up at the normal morning hour at your destination. If your travel plans bring you to your destination well after midnight—say, later than 3 or 4 A.M. local time—you may be better off skipping sleep that night and going to bed the following evening.

Confused? As long as you're flying within 6 time zones, here's a simple way to remember: *Follow the sun*. The sun is in the east in the morning and in the west in the evening. So if you're flying east, get morning light. If you're flying west, get evening light.

Or refer to the following chart:

If you're traveling:	*At destination, begin your light exposure:*	*Avoid light exposure:*
WESTWARD		
2 time zones	2 hours before sunset	Morning
3 time zones	3 hours before sunset	Morning
4 time zones	4 hours before sunset	Morning
... etc.	... to a *maximum* of 6 hours exposure	
EASTWARD		
2 to 6 time zones	Sunrise	Evening of arrival
7 time zones	1 hour after sunrise	Evening of arrival
8 time zones	2 hours after sunrise	Evening of arrival

9 time zones	3 hours after sunrise	Evening of arrival
. . . etc.	. . . continue exposure throughout the day	

Don't expect a complete adaptation on the first day of your trip. Even with these strategies, your body will require a few days to adjust.

Melatonin supplements produce similar effects, but as Dr. Lewy notes, timing is critical. Consult with your physician; if the doctor is unfamiliar with the use of light therapy and melatonin for jet lag, he or she can consult the following medical articles:

- Daan, Serge; Lewy, A. J. Scheduled exposure to daylight: A potential strategy to reduce "jet lag" following transmeridian flight. *Psychopharmacology Bulletin* 1984; 20(3):566–8.
- Lewy, A. J.; Ahmed S.; Latham Jackson, J. M.; Sack, R. L. Melatonin shifts human circadian rhythms according to a phase-response curve. *Chronobiology International* 1992; 9:380–92.

In general, if you have the option it's best to design a melatonin-friendly itinerary that allows you to spend several days in a single time zone, rather than erratically making frequent trips back and forth across multiple time zones.

Shift Work

The same strategies apply for shift work, but they're easier to keep straight. The easiest way to remember is this: *Get exposure to daylight as soon as possible after you're scheduled to wake up, and avoid it before you go to sleep.* So if your bedtime is 10 A.M., don't sit in the sun before you turn in; instead, catch your rays in the evening after you wake up. Similarly, melatonin supplements should be taken shortly before your scheduled bedtime, no matter when it is. In chapter 9, we offer additional strategies to help adjust your biological clock.

CHAPTER 15

Taking Melatonin Supplements

Without a doubt, the best melatonin is the homegrown variety—that is, the melatonin manufactured in your own body. In the previous chapters, we've looked at many ways, large and small, to keep our natural melatonin rhythms strong.

But no matter how well the clock runs, eventually it runs down. Inevitably, as we grow older, our natural wellsprings of melatonin begin to dry up. As you will recall from the first chapter, the rate at which we age is generally set for the entire species. By attending to our melatonin rhythms, we can head off *premature* aging and the early onset of age-related diseases. In short, we can probably stay healthier longer. But when it comes to our genetically programmed life span, nature doesn't grant extensions.

And yet, as we are learning, nature doesn't have the final word. With laboratory animals, experiments have already proved that the genetically programmed aging limit can be overcome by augmenting declining melatonin levels with oral supplements.

For a substance that is changing the way we think about some profound philosophical questions like aging and illness, melatonin is pretty unassuming stuff. Walk into any health food store, and you'll find it for sale. You can buy a month's supply for about ten dollars, and you don't even need a prescription.

Take a capsule in the evening and you may notice that you feel a little drowsier than usual at bedtime. You may dream more than you're used to. Or you may notice no change at all. And yet, if the animal studies translate to humans, you may be adding decades to your life span. That's the potential that melatonin offers, and it's the reason why many people today (including many of the researchers in the field) have started taking melatonin supplements.

That's one view. But looking at the evidence from another perspective, we don't *know* that melatonin will extend human life spans the way it extends those of mice. Humans are by far the most long-lived mammals on the planet, and it may not be possible, even with melatonin's help, to go much beyond what we've achieved with more traditional medical advances. (Even the animal experiments suggest that there is apparently an upper limit to the degree of life extension that can be achieved.)

Further, we haven't been studying the use of melatonin supplements long enough to be able to say for certain that there are no long-term side effects. To be sure, the evidence we have so far points overwhelmingly in that direction: *None of the human or animal studies to date have found any toxic effects from melatonin supplements,* no matter what the dosage. In a world where all medications, from aspirin to Zantac, have some risks, that's a remarkable record. In addition, melatonin is ubiquitous in the body, and it's a natural substance that the body itself creates. And there's no doubt that symptoms of old age are related to declining melatonin levels. But there's always the possibility that melatonin supplements may have some as-yet-unforeseen side effects.

Under certain circumstances, such as jet lag and shift changes, melatonin supplements are an established and proven treatment. But for the larger question of whether they can help us live longer and avoid the physical problems of old age, the jury's still out. Early indications are promising, but the benefits are still unproven.

The only way to answer such questions definitively would be to track the health and longevity of a large group of people taking melatonin supplements over many years. By comparing them with a similar group that didn't take melatonin supplements, we could learn for certain whether melatonin does delay aging and extend life spans in humans. Unfortunately, such tests are unlikely to occur, because of the vast logistical problems they would create.

Options

For most of us, those findings would come too late to help us make a decision. Ultimately, the choice of whether to take daily melatonin supplements is a personal one, based on your judgment of potential benefits and risks. You may wish to start with a short-term course and see if it makes a difference in your health and feelings of well-being.

There are three approaches you can consider. You can rely entirely on lifestyle changes to boost your body's natural production of melatonin. Or you can take daily supplements to boost your body's natural supplies. Or you might choose to take supplements occasionally, at times when you're especially at risk.

How to Know If You Need Supplements

The only way to measure melatonin levels directly is through a blood test, one not available in your doctor's office. In research studies, subjects typically have blood drawn every two hours, since levels fluctuate over the course of the day. That approach, of course, isn't too practical outside the laboratory (and it can't be much fun in the lab, either).

Without a definitive test, the best way to tell whether you need supplements is by assessing troubling symptoms. If you feel tired, if you have trouble sleeping, if you know your lifestyle doesn't give you a lot of exposure to sunlight, you may be suffering from low melatonin levels. You may want to review the checklist on pages 81–83 to determine how melatonin-friendly your lifestyle is.

Taking Daily Supplements

If you decide to take daily supplements, keep in mind that they're only part of the story. Use them to enhance, not replace, your body's own melatonin levels. Your natural rhythms of melatonin—the basic "metronome" of

Risk Factors for Melatonin Deprivation	**Symptoms of Melatonin Deprivation**
Aging	Chronic illness
Exposure to pollution, ultraviolet rays, other free-radical sources	Difficulty concentrating
Intensive exercise	Frequent colds and viruses
Limited exposure to sunlight	Memory disturbances
Certain medications	Mood disturbances (depression, mood swings)
Shift work	Phase-shift symptoms (feeling sleepy during the day, awake at night, etc.)
Smoking	Sleep disturbances
Frequent travel	Unexplained changes in sleep and activity patterns

your body's biological clock—can't be duplicated with supplements. To keep this underlying rhythm strong, implement the lifestyle changes we've already discussed.

Most supplements marketed today contain synthetic melatonin. It's chemically identical to natural melatonin; the only difference is that it's made in a laboratory. Natural melatonin, by contrast, is extracted from the pineal glands of animals—a difficult and expensive process because of the small quantities that the pineal produces. In addition to being less expensive, synthetic melatonin allows greater quality control: It's free of impurities and the potency is standardized. Virtually all melatonin is synthesized in Europe and it's extracted from beans and is white in its pure form. Thus, a good indicator of product purity is color; darker preparations may have other ingredients contained within them.

Melatonin supplements typically come in capsule or tablet form; each capsule usually contains 1 to 5 milligrams. A bottle may contain one or several months' supply. Since the melatonin in capsules is derived from plants, it can be taken by vegetarians and others who avoid animal products. The actual amount of melatonin in each pill is very small, and the capsules generally contain natural, inert fillers to make them easier to handle. You may want to check the label to see what other ingredients they contain. Some preparations combine synthetic melatonin with natural pineal extracts. With these preparations the typical dose is seemingly much higher, on the order of 500 milligrams. However, the actual amount of melatonin that's available to the body is far less than this.

Timing. Make sure you take supplements at the right time. As we age, the pattern of melatonin production by the pineal gland undergoes a very specific change: The *amount and duration* of melatonin produced at night both diminish. The objective with daily supplementation is to restore as closely as possible the melatonin patterns of youth. For that reason, timing is critical.

Early recommendations were to take supplements about half an hour before bedtime. However, individual responses vary; some nutritionists now recommend taking the dose an hour or two before bedtime to give it time to make its way into your system. However, if you find that melatonin makes you drowsy, you can wait until just before bedtime to take it.

In any event, the main point to keep in mind is to take it after darkness falls and before you turn in for the evening. That way, the dose will add an extra pulse of melatonin just as natural levels are peaking. (A timed-release formulation, which would continue to release melatonin all night long, would be ideal. Such formulations are under investigation, but they aren't commercially available yet.)

If you forget to take a dose, simply wait until the following evening. Don't try to "make up" the dose the next morning. It won't provide any benefit; it may make you drowsy, and you may throw off your circadian rhythms.

How much melatonin should you take? There aren't any definitive answers. When prescribing supplements, I usually start with a dosage of 2 to 3 milligrams, increasing the dosage gradually if necessary. That level is high enough to reproduce youthful blood levels of melatonin without causing noticeable side effects. Higher doses may make you drowsy—not a significant drawback, obviously, since you're going to bed anyway. Other than that, there are no reported health problems associated with higher doses of melatonin supplements. In fact, some studies have used doses as high as 40 to 80 milligrams with no adverse effects reported. But by the same token, there's no evidence that higher doses provide any added benefit for most people. For those whose melatonin levels are likely to be extremely low—for example, people in their seventies and older, and people with Alzheimer's disease—a higher dosage may be appropriate. Your doctor may be able to provide some guidance, but since round-the-clock blood tests to measure melatonin levels are impractical except in research studies, the proper dosage for an individual patient can only be estimated.

Another Method: Taking Supplements as Needed

This strategy strikes a middle balance, relying on natural approaches to stimulate melatonin production and the use of supplements for specific or unusual circumstances. If you adopt this method, follow the preceding guidelines on timing and dosage (except in a few instances, which are noted below). Since the use of melatonin supplements to alleviate these conditions is relatively recent, exact dosages have not been established. Generally, a dosage of 2 to 3 milligrams is considered safe and effective.

Here are some specific instances in which you might consider taking melatonin supplements, or increasing your dosage if you're taking daily supplements. Of course, you can also take steps to boost your melatonin levels naturally in these situations.

Travel. Many doctors already recommend melatonin to ease the effects of jet lag. Melatonin may offer added benefits for travelers as well, through its immune-enhancing properties, as travelers are often exposed to new and unfamiliar "bugs." For example, the air in airplane cabins is often loaded with respiratory germs, because of the volume of people enclosed in a small space, the mix of people from diverse regions, and the fact that the cabin air is extensively recycled through the cabin during flight.

Melatonin can help your immune system fight off these unfamiliar invaders. It also can help combat the immune-suppressing effects of travel-related stress. Taking melatonin supplements in the days or weeks before the flight may help prevent a cold from surfacing in the days and weeks afterward.

Shift changes and schedule disruptions. One of the most extensively studied uses of melatonin is to help shift workers adapt to new shifts. Take supplements just before your new bedtime to help reset your circadian rhythms. At the same time, control your exposure to light (see chapter 9), especially during the transitional phase.

You will recall that you can only shift your internal clock a few hours each day. So to make the transition as smooth as possible, shift your bedtime two hours a night over four or five days. If you try to make the shift all on one day, the melatonin won't "entrain" your rhythm, and the transition will probably be more difficult—and take several days anyway.

Even if you don't work shifts, you can use melatonin supplements if you know your schedule will be disrupted. For example, if you're normally in bed by ten but you're working late or making the rounds of holiday parties, try taking supplements around nine-thirty anyway. You may feel a little sleepy at the party, but it will help keep your rhythms on track, and you may feel more rested the next day, even though you got less sleep.

Insomnia. You may wish to take melatonin supplements if you occasionally have trouble getting to sleep. In 1994, a study reported in *The New England Journal of Medicine* confirmed that melatonin supplements can help you fall asleep. Volunteers were placed in a dark room in the middle of the day and instructed to sleep. Those who took melatonin fell asleep within five to six minutes on average, versus a half hour for those receiving placebos. The study didn't look at whether the effects would be the same when melatonin is given at bedtime versus the middle of the day, but there's extensive evidence that melatonin triggers physical changes (such as lowering your body temperature) that prepare the body for sleep.

Stress. At times when you're under a lot of stress, melatonin supplements can help head off trouble. As we saw in earlier chapters, for example, melatonin helps protect the immune system and cardiovascular system from the effects of corticosteroids, which are released from the adrenal glands during times of stress. If you're under increased stress—a deadline at work, a new baby at home—melatonin can help see you through the crisis.

Medications. If you're taking medications that interfere with melatonin production (see chapter 13), melatonin supplements can make up for what your body isn't supplying. That, in turn, can help reduce side effects of these drugs (for example, the insomnia that may result from taking beta-blockers) and help prevent complications.

Infections. During cold and flu season, melatonin can help boost your immune defenses and possibly keep you from catching whatever is going around. Similarly, if your immunity is impaired for other reasons (for example, if you're taking corticosteroids) or you face unusual exposures to germs (such as from working in a hospital), melatonin can help keep you on track.

X rays. X rays and other forms of high-intensity radiation create high levels of free radicals. If you have to be x-rayed, a dose of melatonin beforehand can help protect exposed cells from free-radical damage. (This is one instance where it makes sense to take melatonin supplements during the day.) Take it about a half hour before your visit, so that circulating levels are high at the time of exposure. (You can also take other antioxidants such as vitamins A, C, and E to increase your protection.)

Toxins. Similarly, exposure to toxic chemicals—

pesticides or bleach, for example—can introduce free radicals into your body. If you will be working with or around such chemicals, melatonin supplements can help offset potential damage.

EMF exposure. If you're exposed to high EMF levels, melatonin supplements are probably a good idea. As we've seen, it appears that many (if not all) of the adverse health effects of EMFs are due to their suppression of natural melatonin production. If your EMF exposures at work, school, or home can't be reduced or eliminated, regular melatonin supplements may help mitigate these effects.

Exercise. Exercise and weight training generate free radicals as a by-product of muscular action. In addition, there's evidence that low body fat affects hormone levels—for example, causing the disruption of the menstrual cycle in female athletes who train intensively. Melatonin supplements can help combat the free-radical damage, and may help maintain normal hormonal function as well. Consider taking supplements, especially during periods of intensive exercise—for example, while you're training to run a marathon. (Be careful about timing, though: Take the supplements in the evening, even if you work out in the morning. Otherwise they might make you sleepy.)

Sunburn. Sunburn releases free radicals from damaged cells in the deeper layers of the skin. If you stayed on the beach too long, melatonin supplements may help protect your skin cells against additional damage, and might promote faster healing of the burn.

Other injuries. Other kinds of injuries—burns, abrasions, broken bones, and so on—also generate free radicals. If you're recuperating from an injury, consider taking melatonin to reduce free-radical damage and promote healing.

Where to Get Melatonin Supplements

Melatonin is now widely available in health food stores. Within the health and natural foods industry, some believe that melatonin has more in common with drugs than with nutritional supplements, and should be marketed through traditional pharmaceutical channels. Others, noting melatonin's natural presence in the body, its apparent lack of side effects, and its antioxidant and other health-enhancing properties, maintain that it is more akin to such health store products as vitamin E and beta-carotene.

The same issues are being debated from another perspective by mainstream medical experts. In light of melatonin's benefits and the absence of known risks, many experts see no reason to restrict its availability. In fact, many researchers in the field take daily supplements themselves. Others, however, warn that any substance with such widespread effects on bodily functions must be approached cautiously, especially when we are just beginning to understand its functions and its effects. And they argue that the Food and Drug Administration should regulate melatonin as a drug, and make it available by prescription only.

For now, the easiest way to get melatonin is from your local health food store. However, if it is eventually reclassified as a prescription drug, you'll need to obtain a prescription from your doctor.

If you can't find a local supply, or if you want to find out more specifics about the particular supplement you're taking, here are some leading manufacturers and distributors of melatonin supplements:

European Distributors

Genzyme Pharmaceuticals
Haverhill, Suffolk
United Kingdom CB7 8PU
Telephone: 44 440 703521
Fax: 44 440 707783

Douglas Laboratories
Europe BV
Huls 14
6369 E W Simpelveld
Netherlands
Telephone: 31 45 544 2904
Fax: 31 45 544 5533

U.S. Distributors

Allergy Research Group
400 Preda Street
San Leandro, CA 94577
Telephone: 510-639-4572

Holisticom Systems, Inc.
15 Wabash Avenue
Pittsburgh, PA 15220
Telephone: 412-494-0110

KAL
6415 DeSoto Avenue
Woodland Hills, CA 91365
Telephone: 1-800-755-4525

L & H Vitamins
31–1 Crescent Street
Long Island City, NY 11101
Telephone: 1-800-221-1152
Offers melatonin supplements by mail.

The Pain & Stress Therapy Center
5282 Medical Drive, Suite 160
San Antonio, TX 78229
Telephone: 1-800-669-CALM
Offers melatonin supplements by mail.

Ultra-Vit
236 West Mountain, Suite 105
Pasadena, CA 91103
Telephone: 818-683-8208
Sells through health food stores; no direct sales.

Numerous other manufacturers' and distributors' brands are available in health food stores, including KAL, Only Natural, and Cardiovascular Research.

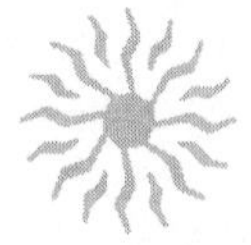

CHAPTER 16

The Future of Melatonin Research

Already, new discoveries about melatonin have dramatically changed our understanding of how the body regulates and restores itself. And the research is only beginning. As scientists learn more about melatonin's wide-ranging effects, they are looking at a growing number of potential ways in which it can be used to prevent and treat disease.

In this chapter, we take a closer look at the vast range of conditions in which melatonin is showing potential as an effective new treatment. Many of these treatments are still in the experimental stage, but others suggest ways in which we can begin, right now, to use melatonin to keep us healthy.

Fighting Environmental Hazards

Melatonin supplements can help protect our bodies from damage by toxins and environmental hazards. In animal studies, melatonin supplements have protected the liver from alcohol-related damage, immune cells from destruction by chemotherapy, and lungs from damage caused by exposure to pesticides. In lab animals, melatonin has prevented cataracts, which are caused by the cornea's exposure to ultraviolet radiation. All of these health problems are related to free-radical damage at the cellular level.

Melatonin as Medicine

Using melatonin to treat disease is appealing for a number of reasons. Most important, it's nontoxic. Melatonin has no known lethal dose; in laboratory experiments, even massive amounts don't cause any harmful effects. Also, it's a natural product, found in all plants and animals. And not least of all, it's inexpensive.

Sleep Disorders

Because melatonin is central to our cycles of sleeping and wakefulness, it's not surprising that one of the most popular uses of melatonin is as a natural and safe sleeping aid. Many people take melatonin in the evening and report that it helps them fall asleep more easily and sleep more soundly. And because these nightly doses of melatonin mimic natural melatonin rhythms, they don't seem to cause the morning grogginess that often occur with traditional sleeping pills.

Medical researchers are also studying the potential of melatonin to help promote a good night's sleep in people whose medical conditions severely disrupt sleep patterns, such as those with Alzheimer's disease and autism.

Heart and Circulatory Disease

Melatonin influences the electrical activity of the heart muscle through the "calcium channel." The calcium channel is a mechanism by which muscle cells "reset" themselves after they've contracted. When muscles contract, they give off calcium, and they can't contract again until they've replenished this supply. The calcium channel allows calcium to flow back quickly across the cell wall, and it determines how quickly and how powerfully the heart can beat. By regulating the calcium channel, certain drugs help the heart beat properly and prevent too-rapid heartbeats. Melatonin also affects this channel, though the details aren't clear yet.

In addition, animal studies show that melatonin protects heart tissue from toxins, presumably because of its antioxidant properties. Melatonin also seems to inhibit clotting in the coronary arteries, and thus reduce the risk of heart attacks.

Melatonin may also head off heart trouble by its influence on the immune system. When the immune system is stressed, the body produces high levels of corticosteroids; these hormones, in turn, raise blood pressure and seem to be a contributing factor in heart failure, heart attack, angina, stroke, and other circulatory problems.

Studies are currently under way to explore the link between the heart and melatonin in more detail.

Obesity

Scattered reports in medical journals speculate that melatonin may play a role in how much we weigh, since it helps regulate body temperature and metabolic rate (that is, the rate at which our bodies consume energy). Studies show that obesity is associated with impaired melatonin secretion. It's possible that as we age, declining melatonin levels make it harder to keep various body systems coordinated with one another and reduce the body's efficiency at converting food into energy.

Conversely, obesity may affect our melatonin levels. The pineal gland produces relatively constant amounts of melatonin from day to day, so as body weight increases, the concentration of melatonin declines. This may explain, at least in part, why excess weight puts us at greater risk for many diseases that are normally associated with old age, including heart disease, stroke, and diabetes.

Glaucoma

Some very recent studies suggest that melatonin may be useful in the treatment of glaucoma by relieving intraocular pressure (that is, pressure inside the eye). However, keep in mind that these findings are preliminary and we don't know all of their implications. For example, by reducing intraocular pressure, it's possible that melatonin supplements could *mask* early symptoms of glaucoma and delay treatment. If you suffer from or are at risk of glaucoma, don't take melatonin supplements without your doctor's knowledge and approval.

Melatonin and the Mind

Seasonal Affective Disorder

Hippocrates wrote, "Of constitutions some are well or ill adapted to summer, others are well or ill adapted to winter." Recent research reveals that the cause is biochemical. Some people are abnormally sensitive to low light levels, and are at risk for severe depression when winter gloom arrives.

Because of the intimate relationship between light exposure and melatonin levels, many researchers believe that the problem lies with the pineal gland. The most powerful evidence of this is the treatment that works best for seasonal affective disorder (SAD). It's believed that for people with SAD, the dim light of winter isn't strong enough to keep the body's daily clock set—in fact, many of those with the disorder show abnormal melatonin rhythms. Exposure to bright artificial light in the morning provides relief in most cases.

SAD is more complicated than a case of the pineal not knowing when to stop, since the symptoms persist even if drugs are given to block melatonin production. Rather, the problem seems to lie with abnormal rhythms of melatonin production.

Postpartum Depression

In women, melatonin levels tend to rise during pregnancy and then drop precipitously shortly after childbirth.

This disruption of normal melatonin rhythms (reinforced, perhaps, by the difficulty of getting a good night's sleep) may contribute to postpartum depression.

Clinical Depression

At least some cases of ordinary clinical depression may be related to melatonin rhythms as well, although, once again, the relationship appears to be complex. For example, many patients with severe depression find their symptoms suddenly lifted if they skip a night's sleep—though the symptoms return once the patient finally falls asleep. In addition, disruptions in sleep rhythms are one of the hallmarks of depression, and those suffering from this disorder often find themselves napping in the daytime and unable to sleep at night. Further evidence of a link between depression and altered melatonin rhythms comes from the fact that many of the medications used to treat depression, including Prozac, stimulate the production of serotonin—the same hormone that the pineal uses to manufacture melatonin.

Epilepsy

Melatonin may help epilepsy treatments work better. Studies have shown abnormal patterns of melatonin production in children with epilepsy. In animal experiments, melatonin has helped stabilize electrical activity in the brain and prevent seizures.

Alzheimer's Disease and Memory Loss

One of the most exciting areas of inquiry right now involves melatonin's effects on Alzheimer's disease. Presently, there's no effective treatment to stop or even slow the progressive memory loss of Alzheimer's, but researchers are hopeful that melatonin supplements may provide some relief and insights into the causes of, and potential treatments for, this debilitating disease.

Patients with Alzheimer's disease have been found to have fewer than normal neurons in the nerve pathway that connects the eyes with the pineal gland, as well as low levels of melatonin. And one of the symptoms of Alzheimer's disease is a loss of circadian rhythms, which add to the patient's sense of disorientation and to the family's or caregiver's burdens. People with Alzheimer's often have erratic sleeping and waking cycles—sometimes sleeping during the day and wandering at night.

This loss of rhythmicity also appears to worsen memory loss by interfering with the brain's ability to sort and file recent experiences in memory. This sorting and filing function typically occurs while we sleep, and the impaired sleep patterns seen in both Alzheimer's and old age contribute to memory loss.

Already some physicians are recommending melatonin to patients with Alzheimer's disease to ease these and other symptoms. In addition, a number of formal studies have been conducted to evaluate melatonin's action in these patients, and additional studies are under way. For example, a small-scale study in Japan found that bright light therapy similar to that used to treat SAD "significantly increased total and nocturnal sleep time and significantly decreased daytime sleep time"—that is, brought sleep patterns for these patients more into line with normal patterns.

Down's Syndrome

In many ways, people with Down's syndrome seem like children who never grow up. And yet physically they grow up all too quickly, exhibiting characteristic signs of aging by the time they reach their twenties: thinning hair, wrinkles, deteriorations in mental and physical faculties.

It now appears that this physical deterioration may be at least partially caused by a poorly functioning pineal gland. The evidence for this includes the following:

- Those with the disorder often have abnormally low or high levels of thyroid hormones. In the laboratory, similar effects are seen in mice who have had their pineal glands removed.
- Sleep patterns in patients with Down's syndrome are often abnormal. In fact, these patterns look similar to the sleep problems that occur in old age, with reduced amounts of rapid-eye-movement (REM) sleep—the dreaming stage of sleep that is essential to memory and other higher brain functions.
- As Down's syndrome patients get older, they're often more prone to infection because of deficiencies in their immune system. This propensity to infection probably occurs because of low levels of melatonin.

Researchers believe that these changes in patients with Down's syndrome may reflect an acceleration of the aging process in the brain and body, and they suggest that oral melatonin may help establish more normal patterns of sleep and activity, thereby improving the quality of life for these patients and their families. In addition, it may be possible to head off immune deficiencies in those with Down's syndrome by starting melatonin treatments in early childhood.

Melatonin and Reproduction

Infertility

In seasonally breeding animals, melatonin plays a key role in regulating the rhythms of reproduction. Through its action on such hormones as estrogen, it also influences the human reproductive system as well, suggesting potential new avenues of research into contraception and fertility. In fact, melatonin seems to cause a mild seasonality in human reproduction (at least in cooler climates), with fertility higher in summer months and a corresponding increase in birthrates during the spring. This pattern has some obvious practical implications for treating infertility, with treatment more likely to succeed in the summer than in the winter.

Contraception

In the Netherlands, some 1,200 women are currently participating in a trial of an experimental birth control pill that combines melatonin and progestin, a female hormone widely used in commercial contraceptives. Early findings show no serious side effects and an effectiveness similar to that of other oral contraceptives. Unlike ordinary contraceptives, however, the pills contain no estrogen, and they may be an alternative for women who can't take estrogen-based contraceptives.

Pregnancy

We've already seen, in the discussion of melatonin and cancer, the ability of melatonin to regulate female hormones. Because of its effects on the reproductive system, melatonin may help reduce symptoms of premenstrual syndrome (PMS) and help prevent problems during pregnancy.

Melatonin plays a big part in pregnancy. In the first twenty weeks, levels of melatonin increase by up to 300 percent. It stimulates production of progesterone, a hormone that helps relax the uterus and helps prevent the immune system from attacking the "foreign" fetus.

In rats, pinealectomy (removal of the pineal gland) increases the rate of miscarriage. In fact, melatonin may one day serve as an early warning system for miscarriage, signaling that a woman may be at added risk if her blood levels of melatonin don't rise as expected. In addition, melatonin research may lead to new treatments to prevent miscarriages in women who are at risk. (A caution, however: We don't know how melatonin affects pregnancy, and the most prudent course is not to take melatonin supplements during pregnancy unless your doctor advises you to.)

Sex

Although the subject hasn't yet been studied in any depth, it's possible that melatonin may play a key role in human sexuality. In seasonally breeding animals, melatonin helps trigger the physical and behavioral changes that bring females into heat and begin courtship and mating rituals. In fact, melatonin has been used to help farmers control when the breeding season begins for swine.

Humans have seasonal breeding patterns as well;

that's why there are so many June babies. These patterns are less pronounced in people, possibly because we tend to live and work in light- and temperature-controlled environments year round. But rhythms are an important factor in sexuality nonetheless. For example, the female reproductive system follows powerful monthly rhythms, which melatonin helps regulate. Daily rhythms also seem to influence sexual expression as well; some people feel more aroused in the mornings, others in the evenings. And, of course, sexuality is closely tied to key events of development and aging across the lifespan, including puberty and menopause.

Though indirect, this evidence suggests that melatonin—and factors affecting it, such as light and EMFs—may influence both sexual drive and fertility. So while it's too soon to offer specific evidence or recommendations, it's possible that strategies for enhancing your melatonin production may give your sex life a boost as well.

Menopause

Osteoporosis—a progressive loss of calcium in the bones that leaves them weak and brittle—affects most women to some degree after menopause, a time during which melatonin levels drop dramatically. Estrogen replacement therapy (ERT) can help prevent osteoporosis, and as we've seen, melatonin and estrogen production are closely linked. Thus, melatonin may help prevent osteoporosis by regulating estrogen levels. Also, animal studies show that the pineal gland has a central role in regulating the metabolism of calcium, and suggests that melatonin supplements and other melatonin-enhancing treatments may help prevent osteoporosis.

Melatonin in Newborns

Infants produce virtually no melatonin in their first three months of life. Some researchers believe that this lag time may be a factor in a variety of newborn conditions.

For example, the immune systems of infants don't begin functioning until about the same time; in their first three months, they're largely protected from infection by antibodies they received from their mothers in the womb.

Also, that age-old bane of newborns and their parents—colic—may be caused by an immature pineal gland, according to a theory advanced in the journal *Medical Hypotheses*. Serotonin—which, as we've seen, is also produced by the pineal gland and is a sort of chemical flip-side to melatonin—tends to increase contractions of the intestines, and may cause the cramping of infant colic. It's possible that colic occurs because these serotonin levels aren't counteracted by melatonin. Significantly, colic usually goes away after the first three months, at just the time when melatonin production gears up.

Too Much of a Good Thing?

Some diseases may occur because the body produces too *much* melatonin, or produces it at the wrong times.

Migraines

Migraines and cluster headaches have cycles of remissions and exacerbations, and symptoms often vary by season. Researcher R. Sandyk, writing in the *International Journal of Neuroscience,* speculates that they may reflect disruptions in the function of the pineal gland. Sandyk reports the successful treatment of one patient's migraine attack with an experimental apparatus that exposed magnetic fields (which, as we've seen, slow or stop the pineal's production of melatonin).

Multiple Sclerosis

The causes of multiple sclerosis (MS) aren't known, but there does seem to be a link between this debilitating disease and melatonin levels. Like diabetes, MS is an autoimmune disorder; its debilitating symptoms are caused when the body attacks the sheaths surrounding the nerves of the central nervous system. Autoimmune disorders represent a failure of the immune system to distinguish between body tissues and foreign invaders, which may be caused, in part, by declining melatonin levels.

Also, MS is an age-related disease. Symptoms most often begin in the late teens or early twenties; many believe the disease may be triggered during puberty, a few years before symptoms become apparent.

Though MS worsens over time, the course isn't straightforward: Symptoms—both physical weakness and an equally debilitating psychological depression—tend to ebb and flow, with periods of physical weakness interspersed with periods in which the body functions almost normally. Significantly, MS symptoms often improve during pregnancy—when melatonin levels are high—and worsen right after childbirth, when these levels fall off dramatically.

In one case, Sandyk found that melatonin levels in a group of MS patients were "significantly lower" than in patients who weren't depressed. Since the psychiatric and physical symptoms of MS often go hand in hand, this finding suggests that low melatonin levels or circadian-rhythm disruptions may worsen MS symptoms, and that melatonin treatments might provide some relief from these symptoms.

However, evidence of the relationship between MS and melatonin is confusing and at times contradictory. Other studies suggest that melatonin may make MS worse. Sandyk reported a "dramatic improvement" in symptoms of a fifty-year-old woman with MS when she was exposed to weak magnetic fields. When the same patient took melatonin, her symptoms worsened. After another round of treatment with magnetic fields, her symptoms improved once again.

Parkinson's Disease

Sandyk found similar success when he used the magnetic treatment for a patient suffering from Parkinson's disease. He suggests that in all these cases, melatonin may act to worsen symptoms.

In these and other areas, scientists are gaining new insights into the broad benefits that melatonin offers, and are seeing exciting possibilities for developing new treatments and enhancing current ones with melatonin.

CHAPTER 17

Beyond Threescore and Ten

On the whole, the Spanish conquistadores were not reflective men, and Ponce de León may not have stopped to think about the broader implications of what might happen if he had actually found the Fountain of Youth.

Those who explore the new worlds of medicine today, however, spend a considerable amount of time wrestling with the potential consequences of their work. The field of life extension is perhaps the area of medicine that raises the most profound questions of who we are and how we live. How, for example, do added years impact the question of *quantity* of life versus *quality* of life? Few of us would be interested in adding more years to our life span if those years are to consist of disorientation, discomfort, and dependence. In fact, a consensus has emerged among physicians and laypeople alike that certain "heroic" measures to extend life—for example, the use of a ventilator to prolong the life of a terminally ill and comatose patient—often don't serve the patient's best interests and may even be immoral. When we speak of life extension, then, we have to mean more than adding years.

In a sense, all of medicine is about life extension. And as insurance life-expectancy tables show, modern medicine has been enormously successful at this effort. Until now, most of the advances have come at the lower end of the scale, through improvements in infant mortality rates and the development of antibiotics to combat infectious disease. More recently, breakthroughs in heart and cancer treatments have helped reduce death and disease among middle-aged populations.

But these advances, dramatic as they are, have only begun to affect the outer end of the scale. Compared with a hundred years ago, many more people are surviving to their seventies and eighties. But once they reach that age, their added life expectancy is about the same as it has always been. In short, modern medicine has made childhood and youth much less dangerous than they used to be, but further improvements in life expectancy will depend, in large part, on medicine's ability to achieve the same results for the last decades of life.

Even so, the advances of modern medicine have already had effects reaching far beyond the doctor's office—some good, some bad, some obvious, some subtle. All of them have implications for life-extension research.

For example, there are some benefits to be gained from an older and more experienced workforce, provided we make changes in the workplace to accommodate them. With people living longer (and needing more money over their lifetimes to support themselves), the tradition of retiring at sixty-five may have to go out the window. It will happen in part by necessity, but also, in part, by choice. Many people will be reluctant to give up their jobs at age sixty-five. Chances are, they will have reached the peak of power, prestige, competence, and compensation. Why walk away from that into a prolonged and premature retirement?

Or you'll see more people, having concluded one career, embark on another one. With financial stability and another ten or twenty years of active work life left, they'll have time, perhaps, to pursue long-deferred dreams in the arts, entrepreneurial ventures, or other goals. And they'll be able to apply a lifetime of accumulated knowledge and experience.

Their employers, coworkers, and employees should benefit as well. When people retire, part of an organization's "collective memory" is lost—the people who can tell you that your latest "new" idea was actually tried back in 1956 and didn't work. There's nobody to tell you how the organization has changed or stayed the same over the years; why it exists; what it believes about itself. There's nobody, even, to tell you that the file you're looking for is in the third drawer from the top, all the way in the back.

And what effects will longevity have on our culture? America, it's said, has a youthful culture. We tend to think of that state of affairs as permanent. We hear that it is rooted in our national consciousness, and has to do with the fact that America is a young New World democracy that stands in counterpoint to the old and stodgy nations of Europe. It's in our blood, we're told. Our fondness for blue jeans and bubble gum, sports cars and rock 'n' roll, is just an expression of this basic national character.

Well, maybe. But rock 'n' roll and blue jeans are big in Europe too. And it's only in the past several decades—since the fifties—that Americans have placed such a high value on youthfulness. Two hundred years ago, people didn't "wash the gray right out of their hair"; they wore powdered wigs that made them look older.

One could argue, instead, that the cult of youth here and abroad has more to do with demographics than destiny. Since the 1950s, the postwar baby boom has created a culture dominated by the images and interests of youth. But already, as the baby boomers now move toward middle age, you can begin to detect a cultural shift, in which the more sober values of middle age and seniority are replacing the exuberance of youth. "Don't trust anyone over thirty" has given way to "Watch out for Generation X."

Even our notions of physical beauty have followed this demographic trend. In the 1960s, who would have predicted that so many popular celebrities and movie heartthrobs would be in their forties or even fifties? Today, middle-aged sex symbols are everywhere. It can't be that these beautiful people just happened to age more gracefully than others from years past; it's also that our cultural standards have changed as the makeup of the population has changed.

What, then, can we expect culturally as a consequence of people living and staying healthy longer? Probably more of the same: an increased emphasis on values that we associate with age. A continuing rise in conservatism. Ads that promote tradition and experience rather than novelty or revolution. In the arts, perhaps a growing appreciation of classic forms and techniques, and less interest in experimentation and the avant-garde.

That doesn't necessarily mean that we'll all be trading in blue jeans for tweed jackets. Nor does it mean that rock 'n' roll is dead. But even these symbols of eternal youth are getting a little gray around the temples.

Of course, there's no way to predict for sure what the world will be like if we succeed in significantly extending the human life span. Undoubtedly, some changes will be positive and some will be negative. But the greatest danger is that we will be unprepared. As you can see, adding years to peoples' lives creates profound changes. The danger is that because these changes come slowly, they will be overlooked, and we won't do what's necessary now to start getting ready for them.

On the other hand, this revolution has been under way for at least a century already. The aging of the population has indeed brought with it unforeseen costs and consequences. But it has also brought vast benefits—a healthier population, more productive people, and, not least of all, countless children who had the opportunity to experience the love and care of older generations.

And in the end, the social impact is of far less consequence than the effects on individual people. All of us know people who would not be here if it weren't for the life-extending advances of modern medicine. Most of us would count ourselves richer for the experience of knowing them.

Sources

Most of the current research about melatonin is described only in scientific journals. As this book was written with the general reader in mind, I have kept references to scientific literature in the text to a minimum. However, some readers may be curious about the recent melatonin studies and may want more detailed information about particular findings. The following are some of the most important accounts of current melatonin research.

Alcohol, Drug Abuse, and Mental Health Administration. 1992. Hickory Dickory Dock/Who's inherited mousie's clock? *The Journal of the American Medical Association* (4):480(1).

Aldeghi, R.; Lissoni, P.; Barni, S.; Ardizzoia, A.; et al. 1994. Low-dose interleukin-2 subcutaneous immunotherapy in association with the pineal hormone melatonin as a first-line therapy in locally advanced or metastatic hepatocellular carcinoma. *Eur J Cancer* 30A(2):167–70.

Bureau, Y. R., and Persinger, M. A. 1992. Geomagnetic activity and enhanced mortality in rats with acute (epileptic) limbic lability. *Int J Biometeorol* 36(4):226–32.

Cagnacci, A.; Elliott, J. A.; and Yen, S. S. 1992. Melatonin: a major regulator of the circadian rhythm of core temperature in humans. *J Clin Endocrinol Metab* 75(2):447–52.

Caroleo, M. C.; Doria, G.; and Nistico, G. 1994. Melatonin restores immunodepression in aged and cyclophosphamide-treated mice. *Ann N Y Acad Sci* 31(719):343–52.

Chen, L. D.; Kumar, P.; Reiter, R. J.; Tan, D. X.; et al. 1994. Melatonin reduces 3H-nitrendipine binding in the heart. *Proc Soc Exp Biol Med* 207(1):34–37.

Chen, L. D.; Kumar, P.; Reiter, R. J.; Tan, D. X.; et al. 1994. Melatonin prevents the suppression of cardiac Ca(2+)-stimulated ATPase activity induced by alloxan. *Am J Physiol* 267(1 Pt 1):E57–62.

Chen, L. D.; Tan, D. X.; Reiter, R. J.; Yaga, K.; et al. 1993. In vivo and in vitro effects of the pineal gland and melatonin on [Ca(2+) + Mg2+]-dependent ATPase in cardiac sarcolemma. *J Pineal Res* 14(4):178–83.

Cooper, K. H. 1994. *Dr. Kenneth Cooper's antioxidant revolution.* Nashville: Thomas Nelson Inc.

Cure for the winter blues. 1987. *Technology Review* 90(8):12(2).

Do you crave light? 1990. *HeartCare* 3(1):16(1).

Dori, D.; Casale, G.; Solerte, S. B.; Fioravanti, M.; et al. 1994. Chrono-neuroendocrinological aspects of physiological aging and senile dementia. *Chronobiologia* 21(1–2):121–26.

Elias, M. 1993. The mysteries of melatonin. *Harvard Health Letter* 18(8):6(3).

Guardiola-Lemaitre, B.; Lenegre, A.; and Porsolt, R. D. 1992. Combined effects of diazepam and melatonin in two tests for anxiolytic activity in the mouse. *Pharmacol Biochem Behav* 41(2):405–8.

Hormone pills and sleep. 1994. *Executive Health's Good Health Report* 30(8):8(1).

Inhibition of melatonin secretion by ethanol in man. 1993. *Metabolism* 42(8):1047–51.

Laakso, M. L.; Leinonen, L.; Hatonen, T.; Alila, A.; and Heiskala, H. 1993. Melatonin, cortisol and body temperature rhythms in Lennox-Gastaut patients with or without circadian rhythm sleep disorders. *J Neurol* 240(7):410–16.

Lesnikov, V. A.; Korneva, E. A.; Dall'ara, A.; and Pierpaoli, W. 1992. The involvement of pineal gland and melatonin in immunity and aging: II. Thyrotropin-releasing hormone and melatonin forestall involution and promote reconstitution of the thymus in anterior hypothalamic area (AHA)-lesioned mice. *Int J Neurosci* 62(1–2):141–53.

Lesnikov, V. A., and Pierpaoli, W. 1994. Pineal cross-transplantation (old-to-young and vice versa) as evidence for an endogenous aging clock. *Ann N Y Acad Sci* 31(719): 456–60.

Light sleepers. 1992. *The Economist* 323(7754):85(2).

Lissoni, P.; Barni, S.; Ardizzoia, A.; Olivini, G.; et al. 1993. Cancer immunotherapy with low-dose interleukin-2 subcutaneous administration: potential efficacy in most solid tumor histotypes by a concomitant treatment with the pineal hormone melatonin. *J Biol Regul Homeost Agents* 7(4):121–25.

Lissoni, P.; Barni, S.; Tancini, G.; Ardizzoia, A.; et al. 1993. Immunotherapy with subcutaneous low-dose interleukin-2 and the pineal indole melatonin as a new effective therapy in advanced cancers of the digestive tract. *Br J Cancer* 67(6):1404–7.

Lissoni, P.; Barni, S.; Cazzaniga, M.; Ardizzoia, A.; et al. 1994. Efficacy of the concomitant administration of the pineal hormone melatonin in cancer immunotherapy with low-dose IL-2 in patients with advanced solid tumors who had progressed on IL-2 alone. *Oncology* 51(4):344–47.

Lissoni, P.; Barni, S.; Tancini, G.; Ardizzoia, A.; et al. 1994. A randomized study with subcutaneous low-dose interleukin-2 alone vs. interleukin-2 plus the pineal neurohormone melatonin in advanced solid neoplasms other than renal cancer and melanoma. *Br J Cancer* 69(1):196–99.

Lissoni, P.; Barni, S.; Tancini, G.; Rovelli, F.; et al. 1993. A study of the mechanisms involved in the immunostimulatory action of the pineal hormone in cancer patients. *Oncology* 50(6):399–402.

Lissoni, P.; Rovelli, F.; Tisi, E.; Ardizzoia, A.; et al. 1992. Endocrine effects of human recombinant interleukin-3 in cancer patients. *Int J Biol Markers* 7(4):230–33.

Lissoni, P.; Brivio, F.; Ardizzoia, A.; Tancini, G.; and Barni, S. 1993. Subcutaneous therapy with low-dose interleukin-2 plus the neurohormone melatonin in metastatic gastric cancer patients with low performance status. *Tumori* 79(6):401–4.

Living in perpetual twilight. 1992. *Nutrition Health Review* Winter(61):5(1).

Maestroni, G. J.; Conti, A.; and Lissoni, P. 1994. Colony-stimulating activity and hematopoietic rescue from cancer chemotherapy compounds are induced by melatonin via endogenous interleukin-4. *Cancer Res* 54(17):4740–43.

Melatonin for jet lag. 1989. *American Family Physician* 40(3):272(1).

Melatonin (editorial). 1993. *British Medical Journal* 307 (6910):952(2).

Melatonin pills for sleep? 1994. *Consumer Reports* 59 (10): 657(1).

Mishima, K.; Okawa, M.; Hishikawa, Y.; Hozumi, S.; et al. 1994. Morning bright light therapy for sleep and behavior disorders in elderly patients with dementia. *Acta Psychiatr Scand* 89(1):1–7.

Mocchegiani, E.; Bulian, D.; Santarelli, L,; Tibaldi, A.; et al. 1994. The immuno-reconstituting effect of melatonin or pineal grafting and its relation to zinc pool in aging mice. *J Neuroimmunol* 53(2):189–201.

Mocchegiani, E.; Bulian, D.; Santarelli, L.; Tibaldi, A.; et al. 1994. The zinc-melatonin interrelationship. A working hypothesis. *Ann N Y Acad Sci* 719:298–307.

Molina-Carballo, A.; Acuna-Castroviejo, D.; Rodriguez-Cabezas, T.; and Munoz-Hoyos, A. 1994. Effects of febrile and epileptic convulsions on daily variations in plasma melatonin concentration in children. *J Pineal Res* 16(1):1–9.

Molina-Carballo, A.; Munoz-Hoyos, A.; Rodriguez-Cabezas, T.; and Acuna-Castroviejo, D. 1994. Day-night variations in melatonin secretion by the pineal gland during febrile and epileptic convulsions in children. *Psychiatry Res* 52(3):273–83.

Morrey, K. M.; McLachlan, J. A.; Serkin, C. D.; and Bakouche, O. 1994. Activation of human monocytes by the pineal hormone melatonin. *J. Immunol* 153(6):2671–80.

Munson, M., and Gutfeld, G. 1993. Bag the lag: pill helps ease fatigue from jet travel. *Prevention* 45(12):24(1).

Murphy, P. J.; Badia, P.; Myers, B. L.; Boecker, M. R.; Wright, K. P., Jr.; and Schlager, D. S. 1994. Early-morning administration of short-acting beta blockers for treatment of winter depression. *Am J Psychiatry* 151(9):1383–85.

National Cancer Institute. 1989. Stage-dependent depression of melatonin in patients with primary breast cancer. *NCI Cancer Weekly,* August 28:19(1).

Neri, B.; Fiorelli, C.; Moroni, F.; Nicita, G.; et al. 1994. Modulation of human lymphoblastoid interferon activity by melatonin in metastatic renal cell carcinoma. A phase II study. *Cancer* 73(12):3015–19.

Nonsteroidal anti-inflammatory drugs affect normal sleep patterns in humans. 1994. *Physiol Behav* 55(6):1063–66.

Pang, C. S.; Brown, G. M.; Tang, P. L.; Cheng, K. M.; and Pang, S. F. 1993. 2-[125I]iodomelatonin binding sites in the lung and heart: a link between the photoperiodic signal, melatonin, and the cardiopulmonary system. *Biol Signals* 2(4):228–36.

Pierpaoli, W. 1993. Pineal grafting and melatonin induce immunocompetence in nude (athymic) mice. *Int J Neurosci* 68(1–2):123–31.

Pierpaoli, W., and Lesnikov, V. A. 1994. The pineal aging clock. Evidence, models, mechanisms, interventions. *Ann N Y Acad Sci* 719:461–73.

Pierpaoli, W., and Regelson, W. 1994. Pineal control of aging: effect of melatonin and pineal grafting on aging mice. *Proc Natl Acad Sci USA* 91(2):787–91.

Pierrefiche, G.; Topall, G.; Courboin, G.; Henriet, I.; and Laborit, H. 1993. Antioxidant activity of melatonin in mice. *Res Commun Chem Pathol Pharmacol* 80(2):211–23.

Pinsky, Mark A. 1995. *The EMF book.* New York: Warner Books.

Puy, H.; Deybach, J. C.; Baudry, P.; Callebert, J.; et al. 1993. Decreased nocturnal plasma melatonin levels in patients with recurrent acute intermittent porphyria attacks. *Life Sci* 53(8): 621–27.

Rojdmark, S.; Wikner, J.; Adner, N.; Andersson, D. E.; Wetterberg, L.; Furuya, Y.; Yamamoto, K.; Kohno, N.; Ku, Y.; and Saitoh, Y. 1994, 5-Fluorouracil attenuates an oncostatic effect of melatonin on estrogen-sensitive human breast cancer cells (MCF7). *Cancer Lett* 81(1):95–98.

Sandyk, R. 1994. Alzheimer's disease: improvement of visual memory and visuoconstructive performance by treatment with picotesla range magnetic fields. *Int J Neurosci* 76(3–4): 185–225.

Sandyk, R.; Tsagas, N.; and Anninos, P. A. 1992. Melatonin as a proconvulsive hormone in humans. *Int J Neurosci* 63(1–2): 125–35.

Sweeney, D. R. 1989. *Overcoming insomnia.* New York: Dutton.

Sze, S. F.; Liu, W. K.; and Ng, T. B. 1993. Stimulation of murine splenocytes by melatonin and methoxytryptamine. *J Neural Transm Gen Sect* 94(2):115–26.

Tan, D.; Reiter, R. J.; Chen, L. D.; Poeggeler, B.; et al. 1994. Both physiological and pharmacological levels of melatonin reduce DNA adduct formation induced by the carcinogen safrole. *Carcinogenesis* 15(2):215–18.

Tohgi, H.; Abe, T.; Takahashi, S.; Kimura, M.; et al. 1992. Concentrations of serotonin and its related substances in the cerebrospinal fluid in patients with Alzheimer type dementia. *Neurosci Lett* 141(1):9–12.

Viviani, S.; Bidoli, P.; Spinazze, S.; Rovelli, F.; and Lissoni, P. 1992. Normalization of the light/dark rhythm of melatonin after prolonged subcutaneous administration of interleukin-2 in advanced small cell lung cancer patients. *J Pineal Res* 12(3):114–17.

West, S. K., and Oosthuizen, J. M. 1992. Melatonin levels are decreased in rheumatoid arthritis. *J Basic Clin Physiol Pharmacol* 3(1):33–40.

Wilson, S. T.; Blask, D. E.; and Lemus-Wilson, A. M. 1992. Melatonin augments the sensitivity of MCF-7 human breast cancer cells to tamoxifen in vitro. *J Clin Endocrinol Metab* 75(2): 669–70.

Index

Page numbers in *italics* indicate illustrations.

D

E

N